Spinoff 1998

National Aeronautics and Space Administration
Office of Aeronautics and Space Transportation Technology
Commercial Programs Division

Developed by
Publications/Graphics Team
NASA Center for AeroSpace Information (CASI)

Foreword

For 40 years, NASA has been an investment in America's future. As explorers, pioneers, and innovators, we boldly expand frontiers in air and space to inspire and serve America and to benefit the quality of life here on Earth. As we enter the 21st century, we will build on these accomplishments with a renewed focus in scientific research and the development and application of new cutting-edge technologies. Our capabilities are unique and will enable us to answer fundamental questions that have challenged humankind for centuries.

Policy and legislative mandates have changed our way of doing business. Through bold initiatives, restructuring, and increased efficiencies, we have reduced our budget and workforce while maintaining a level of excellence that America has learned to expect of us.

In addition, we have established four Strategic Enterprises to implement NASA's mission: Aeronautics and Space Transportation Technology, Space Science, Earth Science, and Human Exploration and Development of Space.

The research and development conducted by our ten Field Centers within these Enterprises will answer many questions about the universe and will result in technologies with wide-ranging applications. The effects will be felt by scientists throughout the world, and by industry in its quest to excel in the highly competitive global economy.

"Spinoff 1998" reflects NASA's versatility and spirit of innovation.

The last four decades of NASA research and development have improved technologies transferrable to industry and impacted our quality of life in very positive ways. Our work in future decades will be even more impressive as we continue to develop technologies that will accrue benefits for all humankind.

Daniel S. Goldin
Administrator
National Aeronautics and Space Administration

Introduction

In 1958, a Congressional Mandate directed the National Aeronautics and Space Agency to ensure for the widest possible dissemination of its research and development results. Thus, the Scientific and Technical Information (STI) Program was born. While this program addressed mostly the timely dissemination of information to NASA, NASA contractors, other government agencies, and the public, technologies were identified that were clearly transferrable and applicable to industry for additional use in the development of commercial products and services. Such considerations spun off the Technology Utilization Program.

The very successful program went through several name changes and is today called the NASA Commercial Technology Program.

The changes that have occurred over time are not only name changes, but program changes that have dramatically altered the philosophy, mission, and goal of the program. It has been identified that a more intense and proactive outreach effort within the program is necessary in order to make the newest and latest technologies available to industry now—at the time the technology is actually developed.

The NASA Commercial Technology Network (NCTN), its interaction with industry at all levels through a large network of organizations and offices, is contributing to the success of small, medium, and large U.S. businesses to remain globally competitive.

At the same time, new products and services derived from the transfer and application of NASA technology benefit everyone.

This publication is one true measurement of NASA's commitment and technique to transfer all applicable technologies through the employment of modern approaches and outreach concepts—the new way of doing business. The benefits (spinoffs) to you, the taxpayer are forever increasing, and improving the quality of life for all humankind. These benefits represent a substantial dividend from the national investment in aerospace research.

We hope that each year this publication continues to demonstrate—through a kaleidoscope of NASA efforts and commercial successes—the variety of projects and programs resulting in transferrable technologies that may be adopted by industry; the mechanism that is in place to facilitate commercialization; and spinoff products and services that have come to our attention in recent times.

Dr. Robert L. Norwood
Director, Commercial Programs Division
National Aeronautics and Space Administration

Contents

Aerospace Research and Development

Whether it was 40 years ago or today—NASA has always delivered. Under the leadership of its Headquarters organization, the field centers were then and are now involved in pioneering and breakthrough R&D efforts.

Each NASA organization contributes to the American scientific and technological growth through the boldness of its programs, and the versatility and uniqueness of its scientists and engineers.

In both aeronautics and space, the results of the many-faceted endeavors have not only ensured a world pre-eminence and dominance for the United States, but also contributed immensely to the benefit of everyone on Earth.

Therefore, many new commercially-developed products and services will be born from the utilization of ambitious NASA research currently conducted in a wide range of disciplines.

NASA HEADQUARTERS

History in the Making

The National Aeronautics and Space Administration (NASA) celebrates forty years of seminal progress. Established in 1958, NASA is steeped in four decades of historical achievement—a twentieth century civilian agency headed to even more impressive accomplishments in the twenty-first century.

Today, the vibrant nature of American aeronautical and space expertise is evident in the air, above the Earth, and outward to distant locales throughout the solar system. New technologies are transforming civil aviation, making air travel safer, more affordable, and less harmful to the environment. The Hubble Space Telescope has become an active, on-duty orbiting observatory that allows astronomers to capture glimpses of new galaxies, view the formation of faraway solar systems, and seek answers as to how the Universe itself began.

Spacecraft are now en route to Saturn and reaching out to asteroid Eros. The moon is being resource mapped, pole to pole, by Lunar Prospector. Mars Global Surveyor circuits the red planet armed with a host of scientific sensors. Meanwhile, the Galileo spacecraft continues on an extended mission to scrutinize Jupiter's ice-covered moon Europa, and that Jovian hot spot of a world, volcanic Io. Lastly, preparations are moving forward to launch the first segments of the International Space Station, an initiative that truly gives humanity a foothold on the future.

These are exciting times. NASA's three-part mission encompasses the embracing of scientific research, space exploration, and technology development and transfer. This effort is built upon a past history of remarkable achievement, but with full knowledge that still more ambitious goals await.

Cold War Winds

NASA was born as a response to Cold War rivalries between the United States and the former Soviet Union, now Russia. The United States suffered a technological slap in the face on October 4, 1957, the result of Russia's launching of the 184-pound Sputnik 1 satellite. As the first artificial object to orbit the Earth, Sputnik 1 ushered in the dawn of what soon became labeled the "space race." The "beep-beep" transmissions from the Earth-circuiting Russian satellite signaled more than an astronautical history-making event. It underscored a battle of ideologies within a world community of nations. Russian intentions in space became even more obvious by the launching of Sputnik 2 just a month later. This satellite was five times heavier than the first Sputnik and carried the first living creature into orbit, a dog named Laika. It became obvious that Russia's space agenda also listed the eventual launching of humans into orbit.

Sparked by the Russian space launchings, America swung into response mode. Congress passed and President Dwight D. Eisenhower signed the National Aeronautics

Astronaut Edwin E. Aldrin, Jr., Lunar Module pilot, poses for a photo beside the deployed United States flag during Apollo 11 extravehicular activity on the lunar surface.

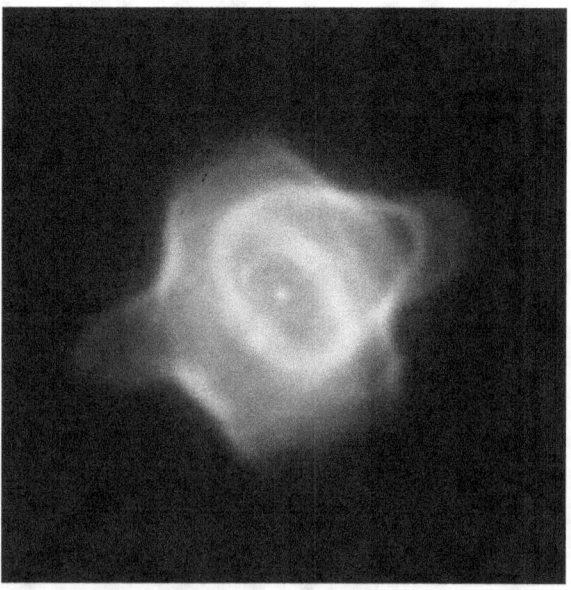

In this Hubble image of the youngest known planetary nebula, the colors shown are actual colors emitted by nitrogen (red), oxygen (green), and hydrogen (blue).

and Space Act of 1958, establishing a new agency with a broad mandate to explore and use space for the benefit "of all mankind." On October 1, 1958, a little less than one year after Russia's Sputnik 1 was hurled spaceward, NASA began its formal work.

In the beginning, the nation's civilian space program was directed by a small cadre of just 170 employees. To add muscle to the newly formed space agency, NASA inherited three major research laboratories from the National Advisory Committee for Aeronautics (NACA): The Langley Aeronautical Laboratory established in Virginia in 1918, the Ames Aeronautical Laboratory set up near San Francisco in 1940, and the Lewis Flight Propulsion Laboratory in Cleveland, Ohio, built in 1941. Additionally, two small test facilities from the NACA were merged into the new NASA, one for high-speed flight research at Muroc Dry Lake in the high desert of California, and one for sounding rockets at Wallops Island, Virginia. Other elements to build NASA were soon added. In December 1958, NASA acquired control of the Pasadena, California-based Jet Propulsion Laboratory, a contractor facility operated for the Army by the California Institute of Technology (Caltech). The first actual NASA installation was the Goddard Space Flight Center, located outside Washington, D.C., in suburban Maryland, staffed by personnel and projects transferred from the Naval Research Laboratory.

Within a short period of time, NASA Headquarters started to orchestrate the country's short-and long-term space agenda. NASA's action plan for shaping both its robotic missions and human space endeavors would rely on establishing solid partnerships between the federal space agency, academia, and the private industry. Under NASA's aegis, numbers of Explorer and Pioneer series spacecraft began charting the space environment, relaying scientific data as to radiation, micro-meteoroid, and solar flare hazards. Information gleaned from these satellites helped in understanding the extent and intensity of radiation belts that were found to surround the Earth.

Significant strides during NASA's early years were made in the satellite applications arena. For example, a 100-foot inflatable satellite, dubbed Echo, was used to reflect radio signals from one point on Earth to another. These experiments and other investigations were forerunners to an explosive growth of satellite telecommunications. Telstar, Early Bird, Relay, and Syncom satellites made possible the commercial satellite operations we routinely enjoy today. These satellites demonstrated the unique attributes of having spacecraft handle telephone, television, and data transmissions around the globe. Along similar avenues, NASA's work in the 1960s with meteorological satellites, specifically Project TIROS (Television and InfraRed Observation Satellite), proved the utility of

Artist concept shows Galileo spacecraft, while still approaching Jupiter. Galileo is flying about 600 miles above Io's volcano-torn surface, twenty times closer than the closest flyby altitude of Voyager in 1979.

watching the world's weather from space. Such images are now considered invaluable for weather forecasting, providing advance warning of violent hurricanes, and El Niño-related events. Likewise, the NASA Landsat satellites, the first of the series lofted in 1972, illustrated how spaceborne sensors could assist in assessing and managing the Earth's precious resources.

Going the Lunar Distance

Nothing helped define NASA's mettle more than human space flight. In April 1959, after a two-month selection process, the space agency introduced to the world seven test pilots as the Mercury astronaut corps. The single-seat Mercury capsule opened the door to U.S.-piloted space exploration.

For NASA to stretch beyond Earth orbit and reach for the Moon, rockets far larger and more powerful were required. In July 1960, NASA's George C. Marshall Space Flight Center was established and charged with rocket booster development. Led by the renowned Wernher von Braun, the giant Saturn V launcher was designed to hurl Americans over cislunar space to land on the crater-pocked Moon.

Providing focus to NASA's technological wherewithal, President John F. Kennedy placed the nation on a lunar trajectory May 5, 1961, asking Congress to support Project Apollo. "I believe this Nation should commit itself to achieving the goal, before this decade is out, of landing a man on the moon and returning him safely to Earth. No single space project in this period will be more impressive to mankind, or more important for the long-range

As America's first experimental space station designed for long duration mission, Skylab program objectives were twofold: To prove that humans could live and work in space for extended periods, and to expand our knowledge of solar astronomy well beyond Earth-based observations.

exploration of space...and none will be so difficult or expensive to accomplish." This visionary call was embraced by the country.

In September 1961, the site of the NASA center dedicated to human space flight would be Houston, Texas. First named the Manned Spacecraft Center, it would later be renamed the Lyndon B. Johnson Space Center in 1973. To test the mammoth Saturn boosters, NASA established in October 1961 the Mississippi Test Facility, an installation later renamed the John C. Stennis Space Center in 1988.

From 1961 into mid-1963, Mercury capsules took astronauts first on suborbital flights, followed by globe-circling orbiting missions. Two-seater Gemini spacecraft flew in 1965 through 1966, giving U.S. astronauts time in orbit to hone skills of rendezvous and docking, procedures necessary to fulfill the objectives of Project Apollo. From data provided by robotic lunar explorers—the Ranger, Lunar Orbiter, and Surveyor moon landers—NASA's Project Apollo led to the first human footprints on the aeon-aged lunar surface on July 20, 1969. In all, a dozen Apollo astronauts would walk across the moon's terrain before the close of 1972. To realize the goal of Apollo, the agency's civil service rolls had grown to 36,000 people. Outside researchers, technicians, and contractor employees that were mobilized to complete Apollo peaked to nearly 377,000 by the mid-1960s. Beyond the technical attributes needed, organizational and managerial competence was marshaled across the country to transform Project Apollo from rhetoric to realization. In a very real sense, footprints embedded in

the moon's dust represent NASA's technological high-water mark for human space exploration beyond Earth orbit.

Defining Events

The momentum of Apollo manifested itself throughout the 1970s and 1980s, with several defining events accomplished by NASA. America's first experimental space station, Skylab, served as a 100-ton home-away-from-home habitat for astronauts. From May 1973 into February 1974, on three separate visits, a trio of astronauts lived aboard Skylab for 28, 59, and 84 days, respectively.

At the height of detente between the United States and the Soviet Union, American and Russian spacecraft linked together as an expression of international space cooperation. The Apollo-Soyuz Test Project brought astronauts and cosmonauts together for the first time, a joint mission that lasted from July 15-24, 1975.

Numbers of NASA interplanetary probes were also launched, such as Mariner spacecraft that studied Mars and Venus on the flyby. Pioneer craft explored giant Jupiter, as well as Venus. Robotic exploration of Mars was capped by the touchdown of Viking landers in 1976. The Voyager 1 and Voyager 2 spacecraft were dispatched in 1977 on trajectories that would yield well over 100,000 up close and personal views of Jupiter, Saturn, Uranus, and Neptune.

Planetary exploration has continued, as evermore capable spacecraft arrive at their destinations. For instance, in 1993, the highly successful Magellan mission to Venus completed three years of mapping nearly all the planet's surface with cloud-piercing radar. The Galileo spacecraft began orbiting Jupiter in December 1995, sending a probe into the murky and complex atmosphere of the gaseous globe. Over the course of the 1990s, the Hubble Space Telescope has relayed striking images of galaxies and other celestial objects, providing a critical new tool to explore the surrounding Universe and all its mysteries. Mars once again has moved center stage in NASA's planetary exploration plans, heralded by the arrival of the Mars Pathfinder in July 1997. Bouncing across Martian terrain on airbags, then coming to rest, the lander unleashed the tiny Sojourner mini-rover. In all, the Pathfinder mission returned more than 1.2 billion bits of computer data and over 10,000 stunning pictures of Mars' surface.

NASA's space shuttle fleet has permitted human access to space since 1981. Scores of missions have given astronaut crews the ability to explore and exploit the microgravity environment of space, then wing their way back to Earth. Versatile in capability, the Space Shuttle program has also enabled flights to Russia's Mir space station—cooperative ventures to gain experience in the building of an international space station. Follow-on space transportation work between NASA and industry is already underway, embodied in such vehicles as the X-33 and X-34 programs and the single-stage-to-orbit VentureStar.

NASA: Future in the Making

Looking forward to the next century, the NASA Strategic Plan for 1998 carries this vision statement: "NASA is an investment in America's future. As explorers, pioneers, and innovators, we boldly expand frontiers in air and space to inspire and serve America and to benefit the quality of life on Earth."

The Plan defines four Strategic Enterprises:

- **Space Science Enterprise** is to solve mysteries of the universe, explore the solar system, discover planets around other stars, search for life beyond Earth; from origins to destiny, chart the evolution of the universe and understand its galaxies, stars, planets, and life;

- **Earth Science Enterprise** is to expand scientific knowledge of the Earth system using NASA's unique vantage points of space, aircraft, and in-situ platforms, creating an international capability to forecast and assess the health of the Earth system; disseminate information about the Earth system; and enable the productive use of Mission to Planet Earth science and technology in the public and private sectors;

- **Human Exploration and Development of Space Enterprise** (HEDS) is to prepare for the conduct of human missions of exploration to planetary and other bodies in the solar system; use the environment of space to expand scientific knowledge; provide safe and affordable human access to space, establish a human presence in space, and share the human experience of being in space; and to enable the commercial development of space and share HEDS knowledge, technologies, and assets that promise to enhance the quality of life on Earth; and the

- **Aeronautics and Space Transportation Technology Enterprise** which has three major technology goals supported by a set of enabling technology objectives. In global civil aviation, a technology goal is to enable U.S. leadership through safer, cleaner, quieter, and more affordable air travel. Another technology goal is to revolutionize air travel and the way in which aircraft are designed, built, and operated. A third technology goal is to enable the full commercial potential of space, and expansion of space research and exploration. This Enterprise also carries a service goal to enable, and as appropriate, provide, on a national basis, world-class aerospace research and development services, including facilities and expertise, and proactively transfer cutting-edge technologies in support of industry and U.S. Government research and development.

With the close of this decade and the turn of the century, the International Space Station is slated to begin operations. Assembly of this huge space facility is an undertaking that involves fifteen countries, led by the United States, Russia, Europe, Japan, and Canada. This complex can serve as a world-class research laboratory. It promises to provide a permanent presence in space and bolsters the prospect that human space travel back to the moon, onward to Mars and other destinations, may occur in the 21st century.

Exploration of the unknown, in aeronautics and space, has been an ongoing assignment of NASA since its creation forty years ago. That quest is to continue in the decades to come. Despite the profound nature and scope of its research, NASA's budget has represented less than one percent of the Federal budget since 1977. For over four decades, NASA's major scientific and technical achievements have contributed not only to an understanding of the origin and development of the Universe, and humanity's place in it—space technologies have been a boon to research in education, transportation, pollution control, rain forest protection, health care, and a host of other practical applications of benefit to all Americans.

By 1975, scientific evidence strongly suggested that Mars had once been a planet with flowing rivers and a denser atmosphere. The first robotic spacecraft to explore Mars—the Viking Landers—were sent to sample the soil and atmosphere to formulate an explanation for its evolutionary past and present state.

AMES RESEARCH CENTER

The strength of American aeronautics owes a great deal of gratitude to the Ames Research Center, situated in Mountain View, California. It was founded in 1939 as an aircraft research laboratory by the National Advisory Committee for Aeronautics (NACA), the committee from which NASA was created. The center is home to three national wind tunnels, including the largest in the world. Ames research in aeronautics is ongoing in fixed-wing and rotor craft, air traffic control technology, artificial intelligence, and human factors.

Ames is NASA's Center of Excellence for information technology (IT). Advanced IT is the vital key to providing revolutionary solutions to the challenges posed by the increasing complexity of NASA's aeronautics and space missions.

Ames' IT effort uses advanced computing systems to analyze data, transforming it into knowledge that can be displayed in visual, virtual, and multimedia environments to aid in the scientific decision-making process. IT systems "learn" as they go, developing the capability to make decisions on the basis of "experience" using limited data inputs. Take, for instance, landing a damaged aircraft safely. Information technologies can draw from a knowl-edge base and make automatic adjustments to the plane, assuring a safe runway touchdown. Aviation operations can also be augmented through IT, providing air traffic controllers, airlines, and pilots with up-to-the-minute information about weather and aircraft position, and will select the best route to a given destination.

Advances in IT will mean intelligent spacecraft can explore planets, comets, and asteroids, working in teams without human intervention. Ames IT specialists envision special software and tiny on-board computers enabling planetary probes to be so small and intelligent that several can be sent on each exploratory mission.

"Our vision in NASA is to open the space frontier. When people think of space, they think of rocket plumes and the Space Shuttle, but the future of space is in information technology. We must develop a virtual presence in space, on planets, in aircraft, and spacecraft," explains NASA Administrator, Daniel Goldin.

Research in aerospace safety at Ames was highlighted in 1997 by creation of a computer generated "virtual" laboratory. The laboratory permits researchers located anywhere in the world to study potentially dangerous aircraft and spacecraft situations without risking human life. In the past, pilots, aerospace engineers, and scientists who were directly involved in tests had to be physically present in a building that houses the world's largest flight simulator.

Ames' simulator is able to move airplane and spaceship cockpits in all directions, including 60 feet vertically and 40 feet horizontally. Five interchangeable cockpits are used to simulate the Space Shuttle, helicopters, airplanes, and other aerospace vehicles. Researchers study aerospace controls, guidance, cockpit displays, automation, and handling qualities of existing or pro-posed aircraft or other vehicles. The simulator creates a convincing environment for a pilot and is controlled by computers programmed to represent each aircraft. Computers calculate correct aircraft response when a pilot changes simulator cockpit controls. In real time, responses by the simulator include cockpit motion, images in the windshield, sounds, and control readouts. Simulations are monitored from control labs at Ames.

Information technology research at Ames is also dedicated to seamless access to resources. Imagine a national computing and information infrastructure that allowed access to the computational resources of the nation in much the same way that one accesses electrical power today. In essence, Ames researchers are at the forefront of creating an "information power grid"—a next generation Internet architecture.

Center experts are also busy defining the prospects for human-centered computing. This work is an effort to build cognitive prostheses, that is, computational systems

The air traffic management software program developed at Ames is used 24 hours a day, 7 days a week for all weather conditions at the Dallas/Fort Worth International Airport—one of the busiest airports in the world.

that leverage and extend human intellectual and perceptual capacities. Human-centered computing is aimed at building computational systems that amplify human intelligence, not substitute for it.

Applying human-centered computing to aviation operations systems is already moving forward. Goals of the work are a major reduction in aircraft accidents and a tripling in the National Airspace System capacity by 2010. To handle the vast amounts of projected air traffic and reduce accidents, computational aids are under design. The envisioned system must not only indicate past and current states of the air traffic control system, but also must be anticipatory of opportunities and risks.

A unique branch of work at Ames is in thermal protection and materials. In conjunction with small companies, the center has been developing new Ultra High Temperature Ceramics—material that will enable sharp leading edges for space vehicles. For Lockheed Martin's X-33 program, a prototype suborbital vehicle to assess single-stage-to-orbit technology, Ames is providing thermal protection system expertise to several selected areas of the experimental craft. Both the Stardust spacecraft that will snag and return comet material to Earth, as well as the Mars Microprobes built to look for the presence of subsurface ice, have counted on Ames' thermal protection system know-how.

NASA's initiative in astrobiology is a primary mission for Ames. Astrobiology is the study of life in the Universe, the story of how an infinitesimal amount of the matter of the Universe assembled into the human mind, allowing humankind to contemplate its history and determine the course of its own evolution. Bringing its interests in astrobiology and information technology together, a NASA Astrobiology Institute is being formed, managed by Ames. This institute is a national consortium of scientists focused on interdisciplinary research, while also training a new generation of researchers with the broad skills, intellect, and enthusiasm to realize the future potential of astrobiology.

Indeed, astrobiology is a broad science effort. It embraces access to space missions, to study stellar nurseries in which planets form and organic molecules are synthesized, to search for life on Mars, to identify habitable planets circling distant stars, and to conduct experiments on adaptation and evolution of life in space. Astrobiology research challenges are profound. A few fundamental questions to ponder: How did life begin? Is there life on other planets, and how can we recognize its presence? How have the Earth and its biosphere evolved and influenced each other over time? What are the prospects for establishing stable ecosystems on Mars to support long-term human presence on that planet?

From early work in aeronautics to advanced computing technology and grappling with the origin of life— Ames Research Center stands ready to discover new worlds, generating new knowledge that stirs the soul, enhances human intellect, and enriches our lives.

Astrobiology is the scientific study of the origin, distribution, and future of life in the universe. This artist's concept represents Mars exploration, which is part of the primary astrobiology mission at Ames.

DRYDEN FLIGHT RESEARCH CENTER

Imagine a desert environment that is just right for good flying weather, on the average of 345 days a year. That ideal situation, and the absence of large population centers, makes the high desert locale of the Dryden Flight Research Center a premier installation for aeronautical flight research. Dryden is NASA's center of excellence for atmospheric flight operations. The center's charter is to develop, verify, and transfer advanced aeronautics, space and related technologies.

NASA Dryden Flight Research Center's F/A-18 Systems Research Aircraft (SRA). Dryden uses the SRA to investigate key technologies, including electrically-powered actuators and computer enhancements that ensure new aerospace concepts are transferred to the U.S. aerospace industry.

Located at Edwards, California, on the western edge of the Mojave Desert, Dryden's history dates back to September 1946. Preparations were then underway to fly the X-1, the first aircraft to fly faster than the speed of sound. That work was partly sponsored by the National Advisory Committee for Aeronautics (NACA), the predecessor organization to NASA.

Since those early days, many milestones in aviation have taken place at Dryden—from supersonic and hypersonic flight, and wingless lifting bodies, to forward-swept wing testing and space shuttle air drops.

Dryden continues to pioneer programs for new aircraft and spacecraft. The center is also aiding the U.S. general aviation community in global economic competition, increasing safety for the flying public.

Just like its past, Dryden remains at the forefront of flight research. One such indicator is Dryden's use of a highly modified F-15 research aircraft. The F-15 is tasked to the Advanced Control Technology for Integrated

Vehicles—or ACTIVE—program. This multi-year flight research effort is improving the performance and maneuverability of future civil and military aircraft flying at subsonic and supersonic speeds. For this program, advanced flight control systems and the ability to thrust vector engine exhaust were integrated into the aircraft.

Other NASA centers have also made use of the ACTIVE F-15 as a flying test bed for experiments. NASA's Lewis Research Center, for instance, is evaluating a computerized system that can sense and respond to high levels of engine inlet airflow distortion or turbulence. This computer system can prevent sudden in-flight engine compressor stalls, potential engine failures, and will also lead to reduced fuel consumption.

High-speed research acoustics is another use of the ACTIVE aircraft. NASA's Langley Research Center used the advanced engine control systems to calibrate engine noise predictions. Such experiments are critical for minimizing noise impact during takeoffs and landings of 21st century High Speed Civil Transport, a second-generation American supersonic jetliner.

The ACTIVE F-15 program is being equipped to assist in the development of advanced "neural network" flight control computer technology. Sponsored by the NASA Ames Research Center, this experiment would allow aircraft control systems to adapt to unforeseen changes in aircraft operating conditions, such as sudden equipment failure.

In 1997, the first flight tests of a "Smart Skin" antenna system took place aboard Dryden's F/A-18 Systems Research Aircraft (SRA). This antenna system may well revolutionize airborne communications. The idea was jointly developed by Northrop-Grumman Corporation and TRW's Avionics Systems Division using internally-generated company funds. The Smart Skin antenna was embedded in a specially-built tip, mounted on the SRA's right vertical stabilizer. Flights of Dryden's F/A-18 substantiated a five-fold increase in voice communication range and a major improvement in the quality of radio transmissions from the aircraft when compared with transmissions from the F/A-18's standard blade antenna.

A Smart Skin antenna has the potential to greatly improve the range and quality of air-to-air and air-to-ground communications. It could also result in improved maintainability and reduced aerodynamic drag. Smart Skin antenna systems could also lead to a 65 percent reduction in airframe structural cutouts for external antennas and a weight savings of 250 to 1,000 pounds per aircraft. Potential application to military needs looks promising, as does use in commercial aircraft. Furthermore, the antenna could be applied to "smart" automobiles and other forms of transportation requiring high-efficiency communications capabilities.

Dryden's F/A-18 SRA was also tasked in 1997 to carry the Advanced L-probe Air Data Integration (ALADIN) experiment. The revolutionary air data probe, designed by Rosemount Aerospace, could result in some informational "magic" for pilots of high-performance aircraft. The new L-probe gives two parameters that standard air data probes do not provide: angle-of-attack (vertical angle of an aircraft's wings and fuselage relative to its actual flight path) and sideslip information (the lateral angle between the aircraft and its actual flight path). The L-probe offers the prospect that the number of probes or vanes needed on a plane can be reduced.

In another milestone, engineers at Dryden completed tests on a device that paves the way for developing future all-electric airplanes that could be safer and more fuel efficient than today's aircraft. Called the Electro-Hydro-static Actuator, the device eliminates or minimizes airborne dependence on pneumatic, hydraulic, and mechanical systems. Flawless performance of the actuator on the left aileron of the F/A-18 SRA was achieved, without using the aircraft's central hydraulics. The actuator and related electrical systems could lead to a five to nine percent fuel savings on an all-electric passenger plane, a thirty to fifty percent reduction in ground equipment, and a reduction in the vulnerability of military aircraft in combat situations.

Other major projects being pioneered at Dryden include the Environmental Research Aircraft and Sensor Technology (ERAST) program. This NASA/industry alliance is expected to lead to a family of unpiloted aircraft that would carry out scientific and environmental missions at heights of up to 100,000 feet. Moreover, such flights can last up to several days or more. ERAST involves a seven-year evaluation program at Dryden that concludes in 2000, designed to rate propulsion, aerodynamics, structures, materials, avionics, and sensor technology used in the remotely controlled aircraft.

A propulsion push into the future is being provided by a Dryden SR-71 aircraft. This high-speed, high-altitude plane has been outfitted with a test model of the Linear Aerospike Rocket Engine. Aerospike engine technology is being refined and developed for space propulsion, and is part of the Lockheed Martin X-33 advanced technology demonstrator for a next generation reusable launch vehicle.

A multi-year hypersonic flight-test program is also underway at the center called Hyper-X. This joint project between Dryden and NASA Langley Research Center is to make use of four unpiloted research aircraft that can fly up to ten times the speed of sound.

Dryden research aircraft are pushing the envelope and providing critical data for NASA's Aeronautics and Space Transportation Enterprise. This work is a commitment to expand human activity and space-based commerce in the frontiers of air and space.

NASA's Advanced Control Technology for Integrated Vehicles (ACTIVE) F-15 aircraft has been highly modified for testing control system technologies, including engine thrust vectoring. Flown by NASA's Dryden Flight Research Center at Edwards Air Force Base, California, the aircraft is heavily instrumented for its research role, including state-of-the-art avionics and cockpit displays.

GODDARD SPACE FLIGHT CENTER

Goddard Space Flight Center is a bustling community of top scientists, engineers, and administrative managers. This NASA center was established in January 1959, named after American rocket pioneer, Robert Goddard, who engineered and launched the world's first liquid-fueled rocket.

As NASA's first major scientific laboratory devoted entirely to the exploration of space, Goddard's mission is in space science, earth science, and technology. Situated in Greenbelt, Maryland, Goddard has the largest scientific staff of all the NASA centers. The center also implements suborbital programs using sounding rockets, balloons, and aircraft from the Wallops Flight Facility on Wallops Island, Virginia.

Theoretical research of the center is carried out at the Goddard Institute for Space Studies. Located in New York City, the institute operates in close association with area universities, and delves into geophysics, astronomy, and meteorology.

Goddard is providing lead support for NASA's Earth Science Enterprise. This endeavor addresses the fundamental question: How can we utilize the knowledge of the Sun, Earth, and other planetary bodies to develop predictive environmental, climate, natural disaster, and natural resource models to help ensure sustainable development, and improve the quality of life on Earth?

To answer this question, a series of satellite launchings have been orchestrated to combine atmospheric, oceanic, and land surface observations into a global environmental study focusing on climate change. Integrated measurements of the Earth's processes from these spacecraft will generate an environmental database focusing on climate change. This unprecedented observational ability can expand our perspective of the global environment to the benefit of everyone.

In August 1997, the Sea-viewing Wide Field-of-view Sensor (SeaWiFS) was launched on the OrbView-2 satellite. This Goddard-managed payload is monitoring global chlorophyll A in the oceans once every two days. SeaWiFS is producing data at a rate unequaled in the history of oceanographic remote sensing, providing scientists with a nearly comprehensive global view of the oceans.

Another part of the quest to study the Earth is measuring rainfall. Rainfall is perhaps the most important factor in defining climate. Excess rainfall can cause flooding and enormous property and crop damage. On the other hand, a lack of rainfall means droughts and crop failure. Rainfall is also a major source of energy that drives the circulation of the atmosphere. Surprisingly, tropical rainfall comprises more than two-thirds of global rainfall.

The Tropical Rainfall Measuring Mission (TRMM), an observatory built by Goddard, was launched into orbit in November 1997. TRMM is a jointly-sponsored project by NASA and the National Space Development Agency (NASDA) of Japan. TRMM is the first mission dedicated to measuring tropical and subtropical rainfall through microwave and visible infrared sensors, and includes the first spaceborne rain radar. TRMM's complement of state-of-the-art instruments will provide more accurate assessments. These new measurements will increase our knowledge of how rainfall releases heat energy to drive atmospheric circulation. Multi-year science data sets

The Earth Observing System (EOS) is the centerpiece of Mission to Planet Earth. EOS will build on the results of past missions, using a series of satellites that will combine atmospheric, oceanic, and land surface observations into a global environment study focusing on climate change.

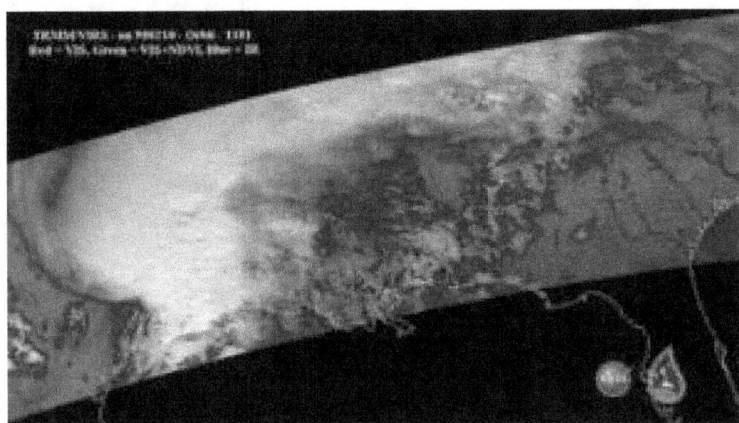

The Tropical Rainfall Measuring Mission (TRMM) is the first mission dedicated to measuring tropical and subtropical rainfall through microwave and visible infrared sensors, and includes the first spaceborne rain radar.

yielded by TRMM will be vastly more informative than any now available.

Additional satellites built for studying the Earth are soon to follow. Slated for launch in 1998, the AM-1 spacecraft is to characterize clouds, aerosol, and the Earth's radiation balance. AM-1 is to cross the equator in the morning hours. The Landsat 7 is set for a 1998 liftoff as well. It will study land surface features and changes using high-resolution imagery. A PM-1 spacecraft is targeted for launching in 2000, geared to study clouds, precipitation and radiative balance. It will cross the equator in the afternoon hours. Still other spacecraft are to be lofted in 2002, 2004, and 2006, each dedicated to studying environmental changes on Earth, both natural and human-induced.

Space-based observing of the Earth is generating a "rain of data"—torrents of information that must be accessed quickly and easily by the scientific research community. Goddard serves as one of several Earth science data centers called a Distributed Active Archive Center, or DAAC for short. The Goddard DAAC's mission is to maximize the investment benefit of the Mission to Planet Earth activities by providing data and services that can help people fully realize the scientific and educational potential of global climate data.

Is the Earth in trouble? Goddard scientists will be gathering significant data gleaned from multiple spacecraft over the years ahead to help answer that question. By learning more about the hazards of global warming, rising sea level, deforestation, ozone depletion, acid rain, and reduction of biodiversity, better stewardship of spaceship Earth is possible.

We now appreciate that the key to gaining better knowledge of the global environment is exploring how the Earth's systems of air, land, water, and life interact with each other. This approach—called Earth System Science—blends together fields like meteorology, oceanography, biology, and atmospheric science. Understanding our changing planet can be expertly done from space. Land cover and land use change, seasonal-to-interannual climate variability, natural hazards research and applications, long-term climate variability, and atmospheric ozone are among the functions best served by spaceborne observation.

Goddard is also responsible for the procurement, development, and verification testing of the Geostationary Operational Environmental Satellites (GOES). NASA also launches the GOES for the National Oceanic and Atmospheric Administration (NOAA).

Not only is the Earth of keen interest to Goddard scientists and engineers. The center has managed the most challenging repair and service missions ever conducted by NASA, the highly successful flights to the Hubble Space Telescope. In February 1997, astronaut teams installed two new instruments and refurbished other elements of the orbiting observatory. In a few years, the Hubble Space Telescope will again be serviced, with Goddard managing the activity. Studies are also underway at the center focused on the Next Generation Space Telescope (NGST), a 21st century near-infrared instrument. NGST will provide major pieces of the puzzle currently missing from the picture of the Universe, its evolution from the Big Bang to the current epoch. This powerful telescope, utilizing a host of advanced technologies, is expected to capture the light from the first stars and galaxies; determine the shape of the Universe and shed light on its eventual fate; map the chemical evolution of the Universe by observing the first supernovae; and observe debris disks around nearby stars.

Detector development at Goddard has enabled trailblazing research in accurately locating mysterious gamma-ray bursts, determining their distance scale, and measuring the physical characteristics of the emission region. Similarly, Goddard engineers have advanced far-infrared detector technology. The center's detector technology has been applied to medical instrumentation, with other commercial applications including industrial manufacturing, environmental monitoring, and agricultural monitoring systems.

Goddard Space Flight Center's vision statement is direct in purpose: "We revolutionize knowledge of the Earth and the Universe through scientific discovery from space to enhance life on Earth."

LYNDON B. JOHNSON SPACE CENTER

In 1998, the greatest peacetime, worldwide collaboration ever attempted will begin building the largest spacecraft in history. The International Space Station is truly a foothold on the future. This orbiting outpost will provide 21st century Earth with unprecedented research advances.

The Lyndon B. Johnson Space Center in Houston, Texas is NASA's designated Center of Excellence for human operations in space. Human exploration and astro-materials is this center's primary mission in support of the Human Exploration and Development of Space (HEDS) Enterprise. And for good reason. Johnson was established in 1961, first as the Manned Spacecraft Center, and later renamed. Decades of milestone-making work first launched astronauts on quick up and down hops, followed by longer and longer Earth orbiting flights, and outward to land on the moon and return to Earth.

Today, teams of American astronauts wing their way into orbit aboard the Space Shuttle Transportation System. Underway since 1981, Space Shuttle flights have placed hundreds of people into orbit and hauled into space several million tons of cargo. The center has management responsibility for the Space Shuttle program. Johnson's Mission Control Center is where all human space flights are monitored. A new, updated Mission Control Center was completed in 1991, and the new flight control room in the new control center is ready to support International Space Station operations as well as Space Shuttle operations.

Selection and training of astronaut crews is done at the Johnson Space Center. Of paramount interest to NASA is understanding the physiological effects of microgravity. Many changes seen in astronauts are still not completely understood. Among them are leaching of minerals from bones, reduction in rate of bone formation, atrophy of muscles when not exercised, and motion sickness. All of the effects of microgravity exposure observed in astronauts have so far been reversed after return to gravity conditions here on Earth.

Numbers of Space Shuttle missions have carried specific experiments to better determine physiological changes in astronauts. Johnson is engaged in an intense and sustained effort aimed at understanding the causes underlying these changes and developing ways to prevent them. The increased information about body functions derived from this effort is paving the way for prolonged missions in space, first aboard the International Space Station, then on sojourns back to the moon and onward to Mars.

In preparation of the construction of the International Space Station, Space Shuttle dockings with the Russian Mir space station returned valuable lessons for extended stays on the high frontier.

Since 1995, U.S. astronauts have resided aboard the Mir orbiting complex. At the close of 1997, the U.S. astronaut time aboard Mir totaled twenty-two months—with eighteen months of continuous occupancy since March 1996. By contrast, it took the U.S. Space Shuttle fleet more than a dozen years and sixty flights to achieve an accumulated year in orbit. Through Shuttle-Mir, NASA has gained valuable experience in rendezvous and docking, spacewalks, and long-duration operation of large-scale systems.

The International Space Station draws upon the resources and the scientific and technological expertise of fifteen cooperating nations, including the United States, Canada, Japan, Russia, and ten member nations of the European Space Agency, and Brazil.

Built in factories worldwide, components for the International Space Station will be lofted into orbit via forty-five launches using three different types of boosters. For five years, an international corps of astronauts will live on and assemble the station in orbit. When completed in 2003, the International Space Station will weigh more than one million pounds. Orbiting at an altitude of 250 statute miles, the huge facility will hold up to seven people when complete.

U.S. orbiter docking with Russian Space Station, Mir.

A science and technology institute in space, an agenda of promising research is to be conducted onboard the International Space Station. Biotechnology, physiology, materials science, combustion science, physics and biology, global environmental observation, and technology development—hundreds of high-quality science and technology experiments are to be conducted year round.

What are the secrets that gravity masks? As gravity factors are eliminated, what are the results? Aboard the International Space Station, the solidification of metals and alloys, as well as transporting fluids and chemicals in microgravity are to be analyzed. Both scientific and commercial benefit is anticipated from such experimentation. Who can say what new materials, more efficient use of fuel resources, new medicines, advanced computers and lasers, and better communications will prosper by International Space Station research?

Engineers, scientists, and astronauts at Johnson are prepared to begin the nation's quest to master space through the International Space Station. Moreover, the U.S. role in the International Space Station maintains the country's leadership position in human spaceflight and aerospace technology.

Demonstrating its vision of the future, Johnson Space Center is testing recyclable life support systems. Volunteer teams have spent extended periods in a three-story, twenty-foot diameter chamber. These teams are investigating the use of mechanical and chemical means to recycle all air and water, including urine. Some of these physico-chemical air and water processors are similar to equipment that will be used onboard the International Space Station. But regenerative life support studies are critical technology for the future of humans in space, because astronauts will not be able to support a trip to Mars or the Moon.

Space engineers at Johnson are also engaged in developing the X-38 prototype space station "lifeboat" or crew return vehicle. To take full advantage of its innovative technology, the X-38 is being designed and tested

The International Space Station is one of the most exciting and challenging international programs leading the world into a new millennium and providing inspiration for future generations.

with an eye toward possible alternative uses as a future international crew transport. Unpiloted X-38 atmospheric test vehicles, built largely at Johnson, are being flight tested in drops from NASA's B-52 aircraft at the Dryden Flight Research Center. An unpiloted space flight test is scheduled for launch aboard a Space Shuttle in 2000. The X-38 is being developed with an unprecedented focus toward efficiency, taking advantage of available equipment and already-developed technology for as much as eighty percent of the spacecraft design.

Johnson Space Center is also the home for research in the fields of life sciences, space systems, robotics, and lunar and planetary geosciences. Indeed, it was a team of Johnson scientists who first found evidence in meteorite ALH84001 that strongly suggests primitive life may have existed on Mars about 3.6 billion years ago. That evidence, while still debated in scientific circles, has fostered considerable research into better understanding the conditions for life, not only on Mars, but on Earth as well.

The X-38 project has two purposes: the first is to build a low-cost crew return vehicle for the International Space Station; the second is to prove that human spacecraft can be built for an order of magnitude of less cost than ever before.

KENNEDY SPACE CENTER

NASA's designated Center of Excellence in launch and payload processing systems is the John F. Kennedy Space Center, situated on Florida's central Atlantic coast. This NASA center has long been a takeoff point for both expendable rockets and space travelers headed for Earth orbit and the Moon.

The Kennedy Space Center grew out of savanna and marsh, prompted by an American commitment in the 1960s to place humans on the Moon through the Apollo program. Kennedy's "space coast" real estate is now the site of Space Shuttle launches. A Shuttle Landing Facility, Orbiter Processing Facility and other installations are in place, built to handle shuttle integration and rollout, payload processing, prelaunch checkout, launch pad operations, launch recovery, and ground turnaround operations.

Kennedy Space Center is placing increasing emphasis on its advanced technology development program. This program encompasses the efforts of the entire Kennedy Space Center team, consisting of government and contractor personnel, working in partnership with

NASA seeks to transfer the KSC-developed Surface Defect Analyzer (SURDA) technology to private industry for use in industrial applications. This system is being developed to provide an accurate, in-field method of evaluating the physical dimensions of surface flaws, defects, and damage on critical surfaces of the Space Shuttle and related ground support equipment.

academic institutions and commercial industry to transfer technology.

This center maintains a vigorous applied research program in support of shuttle launch activities, doing so since 1981. Ground support systems, launch and processing facilities, and environmental protection all receive continued attention for Kennedy Space Center to remain the nation's premier state-of-the-art spaceport.

A hallmark of this center is industrial engineering, typically used to optimize the operations phase of a project or program. The Space Shuttle is NASA's first major program to have a long-term operational phase; however, all the major current and future human space flight programs—the International Space Station, the X-33 experimental reusable launch vehicle, and any lunar base or human space flight to Mars—are also projected to have lengthy operational phases. In this regard, Kennedy has demonstrated a variety of industrial engineering methodologies, categorized into four areas: management support systems, human factors engineering, methods engineering/work measurement, and process analysis and modeling. Each of these areas is producing tangible benefits for NASA and dual-use technologies for other organizations.

Located on the Merritt Island National Wildlife Refuge, Kennedy Space Center workers have always approached their mission with an awareness of the impact on the environment. Kennedy has developed technologies that are environmentally oriented and proactive. Engineers are developing effective methods of cleaning without the use of chlorofluorocarbons. Efforts are also underway that address the safety and disposal of the hazardous fuels used in launch vehicles and satellites. For example, a new scrubber and control system has been devised to eliminate an oxidizer waste stream. The waste stream will be eliminated by using the oxidizer (nitrogen tetroxide) to produce potassium nitrate, a commercial fertilizer, while lowering the oxidizer emissions. Also, an automated multipoint detection system for toxic vapors was designed, originally geared for use on the Space Shuttle launch pad. The FTIR (Fourier Transform Infrared Spectrometer)-based system has been applied to monitoring a wide variety of toxic and contaminant vapors, and would be suitable for many other industrial and government applications.

A range of advanced software programs have been written at the Kennedy Space Center for application to monitoring and diagnostic systems. A ground processing scheduling system provides an artificial intelligence-based tool to aid engineers in scheduling shuttle time and assure that critical tasks are done. Another software advance is represented by a propulsion advisory tool. This expert system focuses on launch day operations to monitor the shuttle's main propulsion system's overall health, following the transfer of liquid hydrogen and oxygen through the ground systems and orbiter into the huge external tank. The software user is warned of potentially hazardous conditions in addition to suggesting a corrective action.

Nondestructive evaluation technology has also been advanced by Kennedy Space Center efforts. Inspection and verification instruments and techniques have been produced. The technology includes, but is not limited to, laser, infrared, microwave, acoustic, structured light, other sensing techniques, and computer and software systems designed to support the inspection tools and methods. This discipline of activity is directed toward reducing shuttle processing costs. Nondestructive evaluation is important for inspecting and verifying space vehicles and their components during manufacture and to continue validating those items during assembly, launch, and on orbit.

The mechanical engineering activities at the Kennedy Space Center have yielded a broad variety of analysis tools, including structural analysis, fracture mechanics, dynamic response, dynamic data, reduction, and processing. Also included are single and multiphase flow, cryogenic fluid flow and storage, and thermal insulation development. Mechanism troubleshooting has been clearly benefited by Kennedy's expertise in this arena.

Advanced electronic technologies that decrease launch vehicle and payload ground processing time and cost, improve process automation, and quality and safety—these are among the accomplishments of Kennedy's Electronics and Instrumentation Technology program. This work has promoted new concepts of data acquisition and transmission, advanced audio systems, digital computer-controlled video, environmental monitoring and gas detection instrumentation, and circuit monitoring instrumentation. The long-term program will develop technology for support of future space vehicles, payloads, and launch systems by advancing the state of the art in launch vehicle and payload processing electronics and instrumentation to reduce costs and enhance safety.

Automation and robotics is yet another aspect of Kennedy Space Center study and application. Payload processing operations present both operational issues and problems. They also provide a forum to take technologies out of the laboratory and make them work reliably in the field. Field testing is critical to the successful insertion of robotic technologies for both NASA and commercial applications. Robotic and automated technology can be applied to a number of ground processing tasks. Kennedy Space Center is working with other NASA centers to develop and apply obstacle-avoidance sensors and systems, multidegree-of-freedom robotic devices, intelligent control systems, inspection sensors and systems, and advanced software technologies for health monitoring and

The Landing Aids Laboratory personnel at Kennedy Space Center has completed the development of a VXI bus-based miniaturized next-generation Microwave Scanning Beam Landing System (MSBLS) Flight Inspection and Certification System. The Space Shuttle uses the MSBLS to provide precision guidance during the last stages of shuttle missions prior to landing.

diagnosis.

Life support for long-term human habitation is the research subject for the Kennedy Space Center's Life Sciences Technology program. Work is underway on the Advanced Life Support (ALS) program. The ALS Breadboard Project is performing biogenerative research and technology development on topics from biomass crop production improvement to resource recovery. The biomass production experiments deal with crop lighting and nutrient-delivery hardware systems, the effects of environmental conditions (i.e., carbon dioxide and temperature) on plants growing in closed chambers, and microgravity effects on plant growth and development. Resource recovery experiments focus on the use of microbiological processes to recycle waste material such as inedible crop biomass into carbon dioxide and mineral forms that can be used by crops and to convert these inedibles into food, thus more efficiently using the limited resources in space, energy, volume, mass, and crew time.

If humankind is literally to reach beyond Earth orbit, ALS research is helping move forward the day when advanced bioregenerative life support systems support crews of astronauts, outward bound to 21st century destinations of asteroids, the Moon and Mars, and ultimately the stars.

LANGLEY RESEARCH CENTER

Research and technology play a vital role in ensuring the safety, environmental compatibility, and productivity of the air transportation system and in enhancing the economic health and national security of the Nation.

Supporting NASA's Aeronautics and Space Transportation Enterprise, Langley Research Center is the Center of Excellence in structures and materials. This center draws upon some 80 years of research, a heritage dating back to 1917 when the center was established.

Langley Research Center is located in Hampton, Virginia and was the first research laboratory for NASA's predecessor agency, the National Advisory Committee for Aeronautics (NACA). The center bore basic responsibility for bolstering the U.S. aviation industry from its earliest beginnings to the position of world leadership that is enjoyed today. From general aviation and cargo-carrying aircraft to hypersonic aircraft and reusable space launchers, Langley is researching, developing, verifying, and transferring advanced aeronautics, space, and related technologies.

A leading example of Langley expertise is embodied within the NASA Hyper-X program, geared to demonstrate supersonic-combustion ramjet (scramjet) technologies. Conducted jointly by Langley and the Dryden Flight Research Center, Edwards, California, the program seeks to demonstrate air-breathing engine technologies. Such an

Hyper-X program features testing of unpiloted aircraft that will fly up to ten times the speed of sound (7,000 mph) to demonstrate "air-breathing" engine technologies.

engine promises to increase payload capacity for future vehicles from hypersonic aircraft to reusable space launchers. By using the oxygen drawn in from the atmosphere—breathing in the air—a scramjet propulsion system permits the discarding of heavy oxygen and associated tanks that rockets must carry for propulsion. NASA has selected a team led by MicroCraft, Inc., Tullahoma, Tennessee, to fabricate a series of small, unpiloted experimental Hyper-X vehicles capable of flying up to ten times the speed of sound. Langley's eight-foot high temperature wind tunnel is on tap to test Hyper-X vehicle designs.

Langley's Thermal Structures Branch has been busy validating a composite intertank structure for the Advanced Space Transportation Program (ASTP). You could say the testing was a snap. The test article was subjected to uniform compression loads to simulate critical load conditions experienced during launch. The 22-by 10-foot part was deliberately broken. For the United States to remain competitive in launching spacecraft, it is necessary to develop a launch system that is lightweight, robust, requires little maintenance or inspection, and has low-cost operations as part of its design features. Langley's ASTP is developing various ways of achieving that goal.

Stepping down in speed, Langley Research Center is also leading a national aviation safety initiative, whose goal is to reduce the aircraft accident rate fivefold within ten years, and tenfold in the next two decades. "Flying already is the safest way to travel. Now it will be even safer," says Jeremiah Creedon, director of NASA Langley. In partnership with the Federal Aviation Administration (FAA), the Department of Defense (DoD), and the aviation industry, research is underway to reduce human error-caused accidents and incidents, predict and prevent mechanical and software malfunctions, and eliminate accidents involving hazardous weather and controlled flight into terrain.

Langley is also working with a consortium led by TRW, Inc., Redondo Beach, California, to demonstrate in-flight a weather-piercing camera that allows researchers to see through fog, smoke, and clouds. The work supports NASA's goal to safely triple capacity at the nation's commercial airports within the next ten years, regardless of conditions that can restrict safe landings and takeoffs of aircraft from airports. The Passive Millimeter Wave Camera project is sponsored by the Defense Advanced Research Projects Agency (DARPA), and is managed by Langley.

In August 1997, Langley researchers demonstrated technology under the Low Visibility Landing and Surface Operations program. NASA's 757 research aircraft was involved in a month-long series of tests that involved computer-generated graphics that outline the correct runway, taxi path, and their precise location on a glass visor mounted between the pilot and cockpit windshield. While taxiing on the airport surface, aircraft position is shown on an electronic moving map on the instrument panel, along with the positions of other active aircraft at the terminal. This activity brought together Langley and Ames Research Center, the FAA, the Volpe National Transportation System Center, and several industry partners.

Langley is helping to advance light plane technologies and revitalize the entire U.S. light plane industry through joint leadership of the Advanced General Aviation Transport Experiments (AGATE) program. This ambitious goal includes working to ensure that light aircraft of the future are dramatically easier to fly, more affordable and safer. For instance, in the safety area, Langley is working with private companies to create airbag technology and energy-absorbing composite structures to protect occupants in small airplanes against fatal injuries. Under AGATE, several airplanes have been crash tested at the Langley Impact Dynamics Facility to demonstrate an improved shoulder harness system and energy-absorbing seats.

Widespread use of composite materials has spurred Langley to sponsor development of Advanced Stitching Machine (ASM) technology under a NASA Advanced Composites Technology program. Partnering with Boeing, an ASM was fabricated that promises to aid in the making of large structures from composites. The goal of program is to make composite wing structures twenty-five percent lighter, to reduce production costs by twenty percent, and to reduce operating costs to airlines. This initiative shows how far aeronautics technology has evolved from the World War II image of "Rosie the Riveter" bolting together segments of metal airplanes.

Langley materials scientists have created a high-performance composite material with a potential market of several billion dollars. In 1997, Langley's PETI-5 was selected in a worldwide competition as one of the 100 most technologically significant new products and processes of that year. This high temperature resin has been selected for use in a U.S. supersonic civil airliner expected to be built early in the next century. The PETI technology has already been transferred to industry with licensing agreements to four different companies. Since

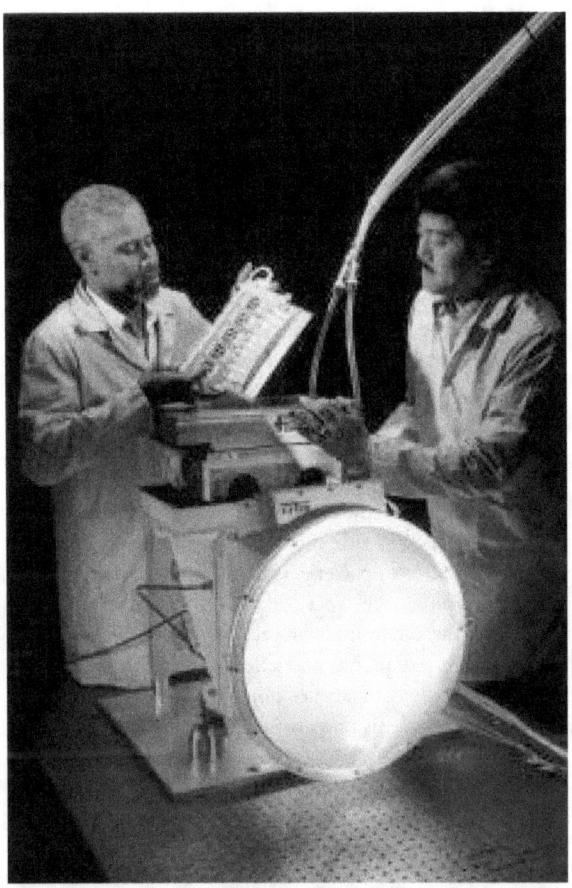

Engineers inspect a passive millimeter wave camera, a weather-piercing camera designed to "see" through fog, clouds, smoke, and dust. NASA Langley is working with a TRW-led industry consortium and the Department of Defense on the project.

currently available metals are either too heavy or cannot withstand the high temperatures created when flying at 2.4 times the speed of sound, composite materials made from graphite fibers and PETI-5 are necessary to both withstand the high temperatures and to make the plane strong enough and light enough to be economically viable.

NASA and industry have teamed to develop the technology necessary to build an economically viable supersonic civil transport, able to carry approximately 300 passengers. This plane—dubbed the High Speed Civil Transport—would halve the flight times from California to Japan, an objective that Langley researchers are confident can be attained. The work is sponsored by the joint High Speed Research Program.

LEWIS RESEARCH CENTER

Lewis Research Center, located in Cleveland, Ohio, is NASA's Center of Excellence in turbomachinery. This center has a primary mission to work in aeropropulsion in support of NASA's Aeronautics and Space Transportation Technology Enterprise.

Organized in 1941 by the National Advisory Committee for Aeronautics (NACA), the forerunner organization to NASA, research efforts at Lewis have impacted every United States aircraft built since the early 1940s. This tradition continues with Lewis engineers and scientists engaged in advancing propulsion technology, allowing aircraft to fly farther, higher, and faster. Research in gaining aircraft fuel economy, noise abatement, and reducing aircraft engine pollution is also underway.

"A dream of wings to come" is a theme actively embraced by Lewis engineers working on the High Speed Research (HSR) Propulsion Project. Early in the next century, American-made supersonic airliners could be carrying millions of passengers per year around the globe at more than twice the speed of sound. Lewis is working on developing the propulsion technology needed to make a new high speed aircraft cost-effective, reliable, and environmentally compatible. This second-generation supersonic airplane is known as the High Speed Civil

Transport (HSCT). To be successful, it must be economically competitive with subsonic aircraft while meeting stringent environmental requirements.

Why a High Speed Civil Transport? World population and economic growth, coupled with lower real travel costs and growing discretionary income, indicate that air traffic is likely to double by 2005. On HSCT international routes—North America to Asia, North America to Europe, and Europe to Asia—demand is expected to grow from 795 billion passenger miles per year in 2005 to more than 2.1 trillion by 2025. High speed travel by HSCT can reduce the travel time to the Far East and Europe by fifty percent within twenty years, and will do so at today's subsonic ticket prices.

One of the more important challenges undertaken at Lewis is HSCT emissions effects on the atmosphere. Under NASA sponsorship, an international group of leading atmospheric scientists are developing atmospheric models which predict ozone depletion to analyze the potential impact of exhaust emissions from a high speed transport fleet. The results to date are very promising. There appear to be areas in the lower stratosphere where future supersonic airliner engines with low emissions could operate without damaging the fragile ozone layer.

Early in the next century, American-made supersonic airliners traveling at twice the speed of sound will carry millions of people around the world. To bring that dream to life, NASA is developing the technology needed to make the High Speed Civil Transport (HSCT) cost effective, reliable, and environmentally compatible.

As part of the HSR program, for which Lewis has responsibility, American industry and government teams have joined forces with Russia's Tupolev Design Bureau. This unprecedented partnership is built upon use of the Russian TU-144 supersonic transport, built in the late 1960s, and withdrawn from passenger service in 1978.

In the TU-144 program, Lewis personnel have worked with researchers from General Electric, Pratt & Whitney, and Russian design teams to plan and conduct two engine ground test experiments. These experiments have provided insight into the inlet/engine interactions which might be expected for an HSCT.

Another exciting research area at Lewis Research Center is the Advanced Communications Technology Satellite (ACTS). This spacecraft was launched aboard the Space Shuttle Discovery in September 1993. Deployed from the shuttle's cargo bay, ACTS has given industry, academia, and government organizations an opportunity to investigate new ways of communicating. In conjunction with industry, Lewis was responsible for developing the ACTS, and is now in charge of management and operation of the unique spacecraft. The ACTS program encompasses an extensive network of ground stations to test and prove pioneering communications concepts and technologies that will advance cheaper, on-demand, flexible communications. Using advanced antenna beams and advanced onboard switching and processing systems, ACTS is pioneering new initiatives in communications satellite technology.

ACTS experiments are showing the way to 21st century applications in telecommunications. The results of the investigations could yield numerous benefits to business, health care, education, national defense, and emergency/disaster relief; and advance the technology in high data rate communications.

Highlighting just a few of the projects under the ACTS Experiments Program would include:

• Increasing the efficiency and lowering the cost of business communications by enabling real-time communications and the use of smaller satellite dishes. ACTS-type technologies can augment fiber-optic networks to extend communications capacity to remote areas, creating new telecommunications users and enhancing the "information superhighway" with Earth/space linkages;

• Transmit images and information to physicians and specialists for use in diagnoses. High-resolution medical imagery from X-rays, MRIs, or CT scans can be sent to another location for review by consulting physicians. The ACTS Mobile Terminal has been used to transmit data from emergency vehicles en route to a hospital.

• ACTS-type technology can provide real-time, more advanced communications capability to the classroom or the workplace. Long-distance, real-time, interactive communications to educate people outside of major learning institutions are being spurred by ACTS demonstrations. Entirely new educational networks could be created.

• The value of advanced military and disaster communications has been clearly reaffirmed since the Persian Gulf war and the aftermath of such disasters like Hurricane Andrew when it battered Florida. Experimenters with ACTS are gaining insight into improved military and emergency/disaster communications by testing new concepts.

ACTS experimenters are exploring a range of voice, video, and digital communications. Investigators include Bellcore, CBS Radio, Stanford Telecommunications, University of Florida, University of Washington, Cray Research, Hughes Network Systems, Sprint, and the U.S. Army Communications Electronics Command.

Lewis Research Center is playing a critical role in assuring American leadership in telecommunications, as well as 21st century aeronautics—two high-technology arenas that will shape national competitiveness in the world community of nations.

The Advanced Communications Technology Satellite (ACTS) provides for the development and flight test of high-risk, advanced communications satellite technology. Using advanced antenna beams and advanced onboard switching and processing systems, ACTS is pioneering new initiatives in communications satellite technology.

MARSHALL SPACE FLIGHT CENTER

To open wide the doors to space development, a new generation of space transportation systems is required. The objective is to provide inexpensive and reliable access to space. An early goal is lowering the cost of placing a pound of payload into Earth orbit from $10,000 to $1,000. And NASA has set a goal to cut space transportation costs by the year 2015 to one percent of today's costs.

Taking on these challenging assignments is the Marshall Space Flight Center in Huntsville, Alabama. The center is NASA's premier organization for developing space transportation and propulsion systems, and for conducting microgravity research. Marshall is NASA's Center of Excellence for space propulsion.

Marshall's Space Transportation Programs Office is developing and demonstrating key, critical technologies to significantly reduce the cost of space transportation. Marshall and industry partners are moving forward on the X-33 program, an effort to demonstrate the key design and operational aspects of a single-stage-to-orbit Reusable Launch Vehicle (RLV) rocket system. Teamwork between government and industry is reducing the risk to the private sector in developing a commercially viable RLV system. In one area of work, the X-33 system is to demonstrate aircraft-like operational attributes, characteristics mandatory to lowering the expense of reaching space.

A second RLV program is the X-34, a reusable technology demonstrator vehicle. The X-34's fast-track development is leading to a vehicle that flies at eight times

the speed of sound and reaches an altitude of 250,000 feet. Marshall is providing design and development of the X-34's main propulsion system, the Fastrac engine.

The first test flights of the X-34 will begin in late 1998, and X-33 flights are slated for mid-1999.

Earth-to-orbit transportation demands highly reusable systems. Marshall's Advanced Reusable Technologies project is pursuing high strength, lightweight structures and cryogenic propellant tanks, durable thermal protection systems, automated checkout and health monitoring of RLV systems, and long-life propulsion components.

Looking even further into the future, Marshall is studying numerous advanced space transportation concepts and technologies. Among these: Air-breathing and pulse-detonation rocket engines and a solar thermal upper stage that uses the Sun's light to produce thrust. Exciting uses of space tethers may one day include generating electrical power for spacecraft, link orbiting satellites to travel in formation, boost satellites to higher orbit, or trail research instruments in the atmosphere.

As stated in the NASA Strategic Plan, "the Human Exploration and Development of Space (HEDS) Enterprise will contribute new scientific knowledge by studying the effects of gravity and the space environment on important biological, chemical, and physical processes. This knowledge will provide fundamental insights for new Earth-bound applications and technology."

Within the NASA HEDS Enterprise, Marshall Space Flight Center's Microgravity Research Program is leading the nation in furthering the development of the space frontier by investigating the fundamental physical, chemical, and biological effects of the microgravity environment of space. Also, the Microgravity Research Program's Space Product and Development Office partners with industry and universities to foster the commercial development of space. Combustion processes are being investigated, as are ways to use microgravity to assess phase changes—when a material changes from one phase—liquid, solid, or gas—to another.

Fluid physics, for instance, is the study of the motion of fluids and the effects of such motion. Since three of the four stages of matter (gas, liquid, and plasma) are fluid, even the fourth, solid, behaves like a fluid under many conditions. Fluid physics is vital, therefore, to understanding, controlling, and improving all of our industrial, as well as natural processes. A low-gravity environment provides scientists near ideal conditions to probe flow phenomena otherwise too complex to study on Earth.

Biotechnology is one discipline that is playing an increasingly important role in medical research and the development of pharmaceutical drugs, agricultural research and products, and environmental protection.

With its elegantly simple, unconventional design, the X-33 is under development with the intention to eliminate any component from the single-stage vehicle that would not be needed later in flight.

The X-34 will demonstrate streamlined management techniques and advanced technologies that have application to future reusable launch vehicle systems. It also may have potential application to commercial launch vehicle capabilities and will provide significantly reduced mission costs for placing small payloads into low Earth orbit.

Major areas of inquiry in this discipline are fundamental to biotechnology science, such as protein crystal growth, and cell and tissue culturing.

A protein crystal growth program has been created to learn how protein crystals grow in space and how to optimize the growth process, while producing large, high-quality crystals of selected proteins. Microgravity conditions inside an orbiting spacecraft, such as the Space Shuttle and the future International Space Station, are relatively free from the gravitational effects of sedimentation and convection. This state of constant free-fall is ideal for studying the mysterious process of crystal growth—what conditions lead to the best crystals and how crystals grow. Improved understanding of the molecular structures and interactions of proteins are important clues that drug designers can utilize to develop new drug treatments that target specific human, animal, and plant diseases. Protein crystal growth experiments and the hardware to investigate crystal growth have both been led by teams of Marshall scientists.

The Microgravity Research Program's work on a bioreactor offers wisdom into how cells endeavor to form complex organisms. Such knowledge is key to understanding the chemistry and mechanics of healthy organs and of cancers, infectious diseases, immune system failures, and other public health problems.

Invented by NASA, the rotating bioreactor spins a fluid medium filled with cells. The spin of the device neutralizes most of gravity's effects and encourages cells to grow in a natural manner. As cells replicate, they "self-associate" to form a complex matrix of collagens, proteins, fibers, and other chemicals. Three-dimensional tissue specimens approximating natural growth are yielded by the bioreactor. These samples provide the opportunity to study the complex order of tissue in a culture system that can be manipulated by drugs, hormones, and genetic

engineering. However, the constant force of gravity here on Earth mechanically limits the size of tissue constructs made in the bioreactor. Early tests of the bioreactor aboard Space Shuttle missions and on Russia's Mir space station point to the growth of much larger, more complex tissue masses than those obtainable in ground-based NASA bioreactors.

Imagine a 21st century future where a space business park is conducting the business of producing made-in-space medicines, speciality glass, and unique electronic components. The first settlements in space, medical research facilities, stop-over tourism hotels—this is a vision attainable by microgravity research carried out today, and work presently in progress to build affordable, reliable, and safe space transportation to open the "Highway to Space."

The NASA bioreactor provides a low turbulence culture environment which promotes the formation of large, three-dimensional cell clusters. NASA-sponsored bioreactor research has been instrumental in helping scientists to better understand normal and cancerous tissue development.

STENNIS SPACE CENTER

The roar of rocketry has long been associated with the John C. Stennis Space Center. Formerly named the Mississippi Test Facility, Stennis is NASA's Center of Excellence in rocket propulsion testing. Located near the Mississippi Gulf Coast, Stennis operations take place within a huge tract of land covering nearly 14,000 acres.

This center is no stranger to the challenging demands of test firing powerful engines. Stennis was tasked in the 1960s and 1970s to test fire all first and second stages of the giant Saturn V rocket used to hurl astronauts to the Moon. In the 1970s, the center took on the role of flight acceptance testing of each Space Shuttle Main Engine (SSME)—a role that Stennis continues today. Every SSME is certified for flight prior to installation in the back end of a shuttle orbiter. Modifications over the years to the SSME, evaluated at Stennis, are making the engine more reliable and less expensive to operate.

Stennis continues its important work in propulsion testing, with the goal of developing a next generation of rocketry that can dramatically drop the cost of placing payloads into space. NASA's Reusable Launch Vehicle work involves testing at Stennis of major elements built for the Lockheed Martin X-33 test vehicle. X-33 cryogenic fuel tanks and components, including the experimental rocket plane's Linear Aerospike Engine, are scheduled to be evaluated and rung-out at Stennis. So, too, is the Evolved Expendable Launch Vehicle (EELV), the Air Force program to drop the cost of launch vehicles from twenty-five to fifty percent. Stennis Space Center's unique test hardware and static test stands will help modernize rocket technology, providing crucial engineering in the creation of new generation reusable and expendable launch vehicles.

While working on technology to propel payloads more cheaply into orbit with greater efficiency, Stennis also has a strong interest in looking at Earth from space, through the tools of remote sensing. Remote sensing is the process of acquiring information about some object or feature of interest from a distance. Data collected by a satellite with a remote sensing device are processed into information. The information that is derived may be used to monitor forest, agricultural, environmental, mineral, or cultural resources. Maps can be created or updated using remote sensing data. Applications for data acquired from remote sensors are quite numerous, with the list growing yearly.

Enhancing U.S. economic competitiveness through development of remote sensing technologies is the mission of the Commercial Remote Sensing Program (CRSP) Office at Stennis. Stennis is the lead center for commercial remote sensing within NASA's Earth Science Enterprise. The center works to assist companies involved in environmental consulting, land use planning, and natural resource management. Stennis is enabling companies to commercialize remote sensing. Through co-funded partnerships, companies use NASA-developed technology to create information products.

One way the CRSP fulfills this objective, is by offering partnership programs that help companies use remote sensing technologies in business applications. This lowers the risk of a commercial group bringing new or improved products and services to market. By working with NASA's CRSP, private concerns can explore remote sensing technologies and equipment without investing large sums of money or revealing company-proprietary information. NASA is a temporary partner. CRSP projects last from three months to three years.

One partnership program is the Earth Observations Commercial Applications Program (EOCAP). Projects under the EOCAP banner are designed for companies already familiar with spatial information technologies. NASA shares technical, financial, and product-development risks with private sector companies while providing access to facilities, technical

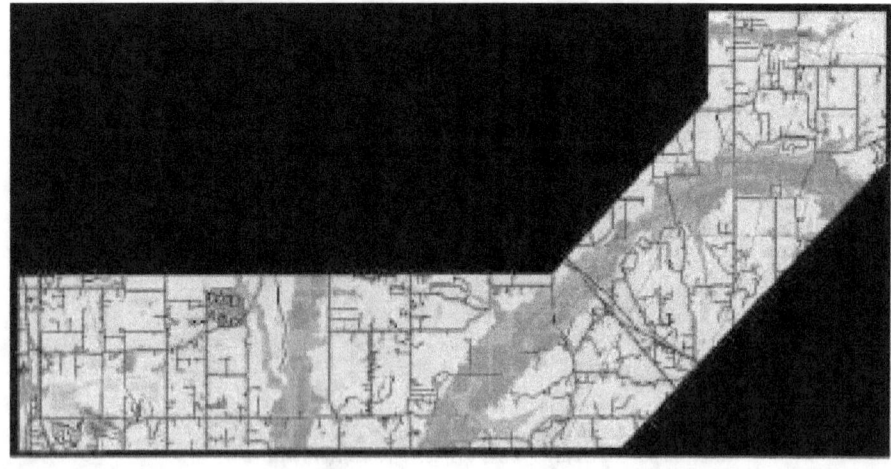

The Commercial Remote Sensing Program recently applied it's comprehensive remote sensing capabilities to highway routing plans for the Mississippi Department of Transportation (MDOT).

expertise, and experienced spatial information specialists. The objectives of the EOCAP partnerships are: To increase the economic benefits from Earth observation and related spatial information technologies; to support the development of new space-based information products and markets; to broaden the application of Earth observation technology within the public and private sectors; and to help create a self-sustaining U.S. spatial information industry.

Recently, two new partnership programs have been announced which highlight advanced technologies. One, the EOCAP-SAR, is focused on synthetic aperture radar (SAR) applications. Unlike conventional remote sensing which requires the sun for illumination, radar images and other useful data, such as ground elevation, can be obtained at night and through clouds.

Another program, the EOCAP-Hyperspectral, is intended to develop applications for remote sensing instruments which produce data in hundreds of spectral colors or "bands" (rather than just a few, such as blue, red, and green). This type of data makes possible vastly improved classification and discrimination of remotely sensed features, but with the concomitant need to store and process enormous quantities of data. Applications of these new types of data are not yet in the mainstream and CRSP intends to bring this technology into common use, much in the same way that digital remote sensing and imaging processing technologies evolved a few short years ago. Looking into the near future, the EOCAP-SAR will help advance SAR technology from the pure science and

research stage to a validated operational, product-oriented status.

The Earth System Science Office (ESSO) at Stennis investigates key biological, chemical, geological, and physical processes, as well as human influences on these processes. This is done through the study of coastal processes in support of NASA's Earth Science Enterprise. Research done at the ESSO has assisted in the fight against the Southern Pine beetle infestation that has struck forests neighboring Stennis. The global carbon cycle of local bays has also been studied, providing new information on shrimp and oyster production.

The ESSO has spurred the use of remote sensing for archaeological and anthropological research. Human adaptation to changing coastal environments can be assessed through remote sensing. Ecological baselines of coastal and estuarial areas are quickly established through the study of archaeological sites. Human settlement subsistence patterns can be examined over changing coastal conditions. In addition, the delineation of ocean processes via remote sensing offers insight into the complex interactions between coastal land and ocean processes, including human influence on those mechanisms.

Stennis Space Center is well-suited to nurture increased application of remote sensing within the public and private sectors. These customers are able to utilize remote sensing data to help businesses grow and give public sector managers more powerful tools to exercise stewardship of the Earth's bounty of natural resources.

The Earth System Science Office (ESSO) at Stennis Space Center has offered ongoing assistance in the fight against the Southern Pine beetle infestation that has struck neighboring territories.

JET PROPULSION LABORATORY

The Jet Propulsion Laboratory (JPL) is the premier center for robotic space exploration. Some of the most profound discoveries about our planetary neighborhood have been relayed from JPL-managed spacecraft dispatched throughout the solar system.

JPL's roots can be traced to the 1930s and 1940s, as students from the Guggenheim Aeronautical Laboratory at the California Institute of Technology in Pasadena pioneered rocket motor technology. This work led to the establishment of JPL. In December 1958, after many years of work for the U.S. Army and other defense agencies, JPL was brought into the NASA organization. Under a continuous NASA contract with the California Institute of Technology, JPL is focused on robotic space exploration for the 21st century.

Peering into the future at JPL is not difficult. Turn any corner of the sprawling complex nestled within the San Gabriel Mountains and you will find cutting-edge

One of several Mars sample return mission concepts under consideration by NASA and engineers at Jet Propulsion Laboratory.

research. JPL's Center for Space Microelectronics Technology (CSMT) is a case in point.

CSMT's Microdevices Laboratory is a state-of-the-art facility focused on creating the building blocks enabling NASA's vision of smaller, faster, cheaper spacecraft. Micromachined seismometers, gyroscopes, accelerometers, even weather stations, are being designed at CSMT. Devices based on silicon, III-V compound semiconductors, and superconductors can be fabricated with nanometer-sized features. A multitude of advanced microdevices are being developed, such as infrared detectors, millimeter and submillimeter wave sensors, ultraviolet, x-ray, and photonic devices, micromagnetic devices, chemical sensors, and electronic neural networks. Rapid prototyping of devices—from concept to actual technology—in less than two years has been demonstrated.

Returning information over laser beams from far away distances across the solar system is another promising research arena. Laser communication (lasercom) technology has been under active JPL development for several years. Laser satellite communications terminals would benefit civilian, military, and commercial interests. This next generation of telecommunications could replace everything from fiber/cable links and microwave to traditional radio frequency communications. Commercial applications include communication between distant buildings, as well as bypassing normal Internet backbone hardware through the use of communicating satellites ringing the Earth.

Engineers in JPL's Optical Communications Group have shown that laser satellite communication equipment can provide advantages of 3:1 in mass and 2:1 in power relative to microwave systems. The vast distances to deep space make data return via conventional radio frequency techniques extremely difficult. Lasercom technology can meet the needs of a variety of space missions, including intersatellite links, Earth to near-space links, and deep space missions.

Microdevices and laser telecommunications represent a tiny fraction of JPL's roadmap into the future. JPL is the Center of Excellence in deep space systems, and provides instrument technology for NASA's Earth Science Enterprise. In this regard, JPL scientists have built an integrated program of research on the El Niño-Southern Oscillation phenomenon. El Niño monitoring via satellites has helped validate methods to better understand and appreciate the nuances of global climate change.

JPL's New Millennium program features smaller, more compact and more versatile spacecraft than robotic probes of the past. Carrying state-of-the-art electronics, propulsion, sensors, and other hardware, New Millennium craft are being built to fly to asteroids, comets, and develop the technology required to search for planets circling other stars. The first of this class of spacecraft is Deep Space 1 (DS 1). A primary objective of DS 1 is to validate solar electric propulsion, among twelve new technologies being evaluated for use on future space missions.

In development at JPL are missions to explore Mars, Jupiter's moon Europa, as well as the Sun and distant Pluto. The red planet Mars is the site for increased robotic exploration as Mars Surveyor orbiters and surface landers reconnoiter the planet. These missions are designed to understand Martian geology, geophysics, mineralogy, and climate, helping to determine whether or not Mars has been, or is even now, an abode for life. Critical to adequately survey Mars and its range of geological diversity is mobility. Robotic vehicles capable of rolling across the Martian terrain are required, not only to move from locale to locale, but also to inspect and gather soil and rock samples. Part of the challenge will be to establish criteria to distinguish between materials of biological and non-biological origin both for sample selection and in sample analysis on Earth. In 2005, the first Mars sample return mission will be underway.

Another JPL mission being readied is Stardust, which begins a trek to comet Wild-2 in February 1999, collecting dust and other materials tossed off from the object, then returns those samples to Earth.

A trio of JPL missions, tagged "Ice and Fire" with spacecraft launched in 2003, 2004, and 2007, respectively, are headed for ice-covered Europa, distant Pluto, and to fly close-up to the Sun.

JPL's exploration quest includes the Origins program, a sweeping initiative within the NASA Space Science Enterprise to address how the universe, galaxies, stars, and planets form and evolve. A sequence of JPL projects is

In the past, all new instruments have made their debuts on expensive missions, but that is about to change. The New Millennium program is a series of low-cost space missions expressly designed to test out new technologies, starting with the DS 1 mission in 1998. This program is but one example of how JPL works with its sponsors to find creative solutions for the successful development of innovative and new technologies.

being blueprinted to first detect, then image and survey, Earth-like planets beyond our solar system. These JPL pursuits call for the development and utilization of revolutionary technologies to achieve mission goals considered impossible in prior decades. Demonstrating the challenges ahead is JPL's Space Interferometry Mission (SIM). It will be the world's first long-baseline optical interferometer in space. With the unprecedented astronomical accuracy and high spatial resolution, SIM is being designed to allow indirect detection of planets through observation of thousands of stars and investigate the structure of planetary disks with nulling imaging.

SIM is the technological precursor to the Terrestrial Planet Finder, an infrared interferometer assigned the difficult duty of direct detection of terrestrial planetary companions to other stars, and also for detecting spectral lines which might indicate a habitable planet. If successful, 21st century technology will allow the spotting of "pale blue dots." These Earth-like worlds may be the future destination of robotic interstellar probes, followed by star sailors of generations hence.

TOPEX/Poseidon satellite measures the precise shape of the ocean's surface and how it changes through time. The satellite's measurements are the most precise tool we have for figuring out if sea level is rising, calculating ocean currents, and identifying climate trends such as El Niño.

Technology Transfer and Commercialization

Forty years ago—in July of 1958—a congressional mandate directed the National Aeronautics and Space Administration to ensure the widest possible dissemination of information resulting from its R&D efforts.

This gave birth in, 1962, to NASA's Technology Utilization Program. It included three Industrial Assistance Centers, released the first Tech Brief in loose-leaf format, expanded the industrial outreach by increasing the Assistance Centers to ten over a period of years, and started COSMIC, the Computer Software and Management Information Center.

In recent years, the Program has changed its structure, goal, and mission, and has broadened its scope. It is currently known as the NASA Commercial Technology Program, with a wide network of organizations ready to do business in a different, improved, and more appropriate way.

Today, the NASA Commercial Technology Network provides a great number of different publications and services geared to enhance and further the global competitiveness of U.S. industry. The Commercial Technology Division at NASA Headquarters and the Commercial Technology Program offices at the ten field centers serve as gateways to accessing the cutting-edge research and technology available for transfer and commercial use. Each year, NASA Spinoff highlights the activities and successes of one of the agency's field centers. This year, the spotlight will focus on Dryden Flight Research Center.

Spotlight on Dryden Flight Research Center

Many of the aircraft at Dryden Flight Research Center, like these pictured on the runway, have led to major advancements in the design and capabilities of many civilian and military aircraft. Dryden's Technology Transfer and Commercialization Office assists in the transfer of this cutting-edge research and technology to the U.S. aircraft industry.

In March 1994, Dryden Flight Research Center became a full-fledged field center. As a result of attaining center status, Dryden's Technology Transfer and Commercialization (T2) office was officially created in May 1995. The T2 office is now part of the "PACE" group, which consists of Public Affairs, Commercialization, Education, and History.

As NASA's primary installation for flight research over the last fifty years, Dryden projects have led to major advancements in the design and capabilities of many civilian and military aircraft. The T2 staff work closely with a variety of clients, from large aircraft companies to small business owners from the start, to ensure that the technologies tested and developed at the center can be directly integrated by its customers.

The T2 staff developed their own guidelines and work closely with the researchers and engineers at the center to capture Dryden's unique innovations. Aeronautics research at the center results in technical papers, the essence of which are captured in NASA's TechTracS database. The T2 staff also work in tandem with the Research Engineering, Projects and Facilities directorates to determine which innovations should be protected by patents.

In addition, the staff work closely with contracting managers' technical representatives to determine what technologies have been developed as a result of contracts, grants, and Small Business Innovation Research (SBIR) contracts. This cooperation resulted in many new patents and a faster transfer of new technologies to U.S. industry. To facilitate new technology reporting, the T2 staff encourage the companies to write articles for *NASA Tech Briefs,* and follow up with press releases to help launch the companies' new innovations.

Striving to meet the aggressive goals set forth by Administrator Dan Goldin in NASA's Strategic Vision, the following projects and partnerships provide an overview of Dryden's role in meeting NASA's goals. This roadmap for the future is outlined and best illustrated by the "Aeronautics and Space Transportation Technology: Three Pillars for Success."

Pillar One: Global Civil Aviation

To preserve the nation's economic health and the welfare of the traveling public, the first pillar focuses on safety, environmental compatibility, and affordable air travel. Dryden's first goal is to reduce the aircraft accident rate by a factor of five within ten years and by a factor of ten within twenty years. To meet the challenge, current innovations and programs include the patented Propulsion-Controlled Aircraft, the Structural Health Monitoring System on Systems Research Aircraft for X-33, and the Lidar Clear Air Turbulence Measurement, developed under an SBIR contract by Coherent Technologies.

The innovative work of reducing emissions of future aircraft by a factor of three within ten years, and five within twenty years, is ongoing. In addition, a new family of remotely-piloted vehicles (RPVs) that fly slower, higher, and longer are being developed. These long-duration, high-altitude RPVs could be used in upper-atmosphere science missions to help collect, identify, and monitor environmental data to assess global change. They could also carry telecommunications equipment to high altitudes, serving much like satellites for a fraction of the cost of putting traditional satellites in space.

The Environmental Research Aircraft and Sensor Technology (ERAST) partnership has yielded significant technological advancements to meet the second goal of the first pillar—environmental compatibility. Besides Dryden, three other NASA research centers (Ames, Langley, and Lewis), along with The Association for Unmanned Vehicle Systems International, American Technology Initiatives Inc., Thermo Mechanical Systems Inc., and several universities are included as partners. The major industry partners are listed below in parentheses, after the names of the unpiloted aircraft, which are products of the ERAST partnership.

- Pathfinder (AeroVironment Inc.) is a proof-of-concept vehicle for two more prototype solar-powered aircraft that have the ability to study the upper atmosphere without disturbing it. Remote monitoring of storm developments, forests, and crop damage are also benefits.
- Apex (Advanced Soaring Concepts, Inc.) is used to validate high-altitude testbed aircraft design methodologies by measuring airfoil characteristics at low Reynolds numbers and high subsonic Mach numbers in a low-turbulence environment.
- Perseus B (Aurora Flight Sciences Corp.) involves test engine concepts, lightweight structures, science payload integration, and fault tolerant flight control systems.
- Altus (General Atomics Aeronautical System Inc.) verifies technologies that lead to a long-duration, high-altitude vehicle that could carry science payloads.
- Demonstrator-2 (Scaled Composites Inc.) centers on over-the-horizon communication capabilities, lightweight structures, science payload integration, engine development, and flight control systems.

The second goal is to reduce the perceived noise levels of future aircraft by a factor of two from today's subsonic aircraft within ten years, and by a factor of four within twenty years. Dryden is currently flying the SR-71 aircraft, on loan from the U.S. Air Force, to study sonic

(Continued)

NASA, industry, and universities are working together as Environmental Research Aircraft and Sensor Technology (ERAST) partners to develop a new family of remotely-piloted vehicles that will fly slower, higher, and longer than ever before.

Spotlight on Dryden Flight Research Center *(Continued)*

boom propagation. Data from the SR-71 high speed research program will be used to aid designers of future supersonic/hypersonic aircraft and propulsion systems, including a high speed civil transport.

The final goal is affordable air travel. While maintaining safety, Dryden strives to triple the aviation system throughout, in all weather conditions within ten years. Reducing the cost of air travel by twenty-five percent within ten years and by fifty percent within twenty years, is being accomplished by Orbital Sciences' patented Adaptive Performance Optimization, Electro Hydrostatic Actuator, the Fiber Optic Position Measurement Network, and the Smart Actuator, which reduces wire weight by sixty percent and detects its own internal failures; thereby reducing maintenance costs.

Demonstrator-2, one of the projects within the ERAST program, was built by Scaled Composites Inc., and is testing technologies that could result in long-duration (12 to 72 hours), high-altitude vehicles capable of carrying science payloads. Lightweight structures, science payload integration, engine development and flight-control systems are key technology development areas for th is project.

Pillar Two: Revolutionary Technology Leaps

NASA's objective outlined in the second pillar is to explore high-risk technology areas that can revolutionize air travel and create new markets for U.S. industry. The technology challenges for NASA include: eliminating the barriers to affordable supersonic travel, expanding general aviation, and accelerating the application of technology advances.

The first technology goal is to reduce the travel time to the Far East and Europe by fifty percent within twenty years, and to do so at today's subsonic ticket prices. To break the barriers to high speed travel, Dryden has imitated several programs including the TU-1 44LL Initiative.

The Russian TU-144LL supersonic transport is being used as a flight research test vehicle to conduct experiments which will enhance the development of advanced technology necessary to build the next generation high speed civil transport. The purpose of the program is threefold: develop and transfer supersonic airliner technology directly to U.S. aircraft industry under NASA sponsorship; establish a working relationship between U.S. and former Russian aircraft manufacturers at

programmatic and technical levels; and acquire flight research data and existing operational experience to narrow the margin of European supersonic transport experience gained from the Concorde.

Dryden is also dedicated to providing next-generation design tools and experimental aircraft to increase design confidence and cut the development cycle in half. This goal is being accomplished through several programs including: the Hyper-X program, which will demonstrate hypersonic propulsion technologies; the X-36 Tailless Fighter Agility Research Aircraft, a remotely-piloted jet designed to fly without traditional tail surfaces aimed at improving the maneuverability and survivability of future fighter aircraft; and the Linear Aerospike SR-71 Experiment (LASRE), designed to gather data on the aerospike's exhaust plume as it travels through the transonic region of flight (just below to just above Mach 1).

Pillar Three: Access to Space

Finally, low-cost space access is essential to unleashing the commercial potential of space and greatly expanding space research and integration. NASA's primary space transportation technology role is to develop and demonstrate pre-competitive, next-generation technology that will enable the commercial launch industry to develop full-scale, highly competitive, and reliable space launchers.

Dryden is pushing ahead with two programs with objectives to meet the goal of this last pillar. First, in partnership with Lockheed-Martin in the X-33 program, Dryden aims to reduce the payload cost to low-Earth orbit by an order of magnitude, from $10,000 per pound within ten years. Second, in partnership with Kelly Aerospace, the Eclipse program goal is to reduce the payload cost to low-Earth orbit by an additional order of magnitude, from thousands to hundreds of dollars per pound, by the year 2020.

The flight research programs initiated at Dryden help move the U.S. aircraft industry closer to achieving its goals. And everyday the efforts of the Technology Transfer and Commercialization office take Dryden Flight Research Center one step closer to realizing its vision statement: "The world leader in flight research for discovery, technology development, and technology transfer for U.S. aeronautics and space preeminence."

Dryden is flight testing the Linear Aerospike SR-71 Experiment (LASRE), which is designed to gather data on the aerospike's exhaust plume as it travels through the transonic region of flight (just below to just above Mach 1).

Technology Transfer & Commercialization

Throughout the nation, the NASA Commercial Technology Program sponsors several centers and their activities which are designed to assist U.S. business and industry in accessing, assessing, utilizing, and commercializing NASA-sponsored technology. These organizations work closely with NASA to provide a full range of technology transfer and commercialization services and assistance.

The hub of the network is the **National Technology Transfer Center** (NTTC) *<http://www.nttc.edu>*, located at Wheeling Jesuit College in Wheeling, West Virginia. The NTTC is an integrated resource for accessing federally-funded research and development and other information. By way of the NTTC Gateway, a free service, the private sector

The Ames Technology Commercialization Center (ATCC) is located in San Jose, California. As a business incubator, the ATCC provides office space and related business services for local start-up companies utilizing NASA technologies.

can maintain person-to-person contacts in the federal laboratory system. The Gateway provides a direct link to many resources including technical reports, technologies, and facilities at NASA and other federal laboratories.

The NTTC is also responsible for administering **NASA TechTracS** *<http://ntas.techtracs.org>*, which provides access to NASA's technology inventory and numerous examples of the successful transfer and commercialization of NASA-sponsored technology. TechFinder, the main feature of the Internet site, allows users to search the technologies and success stories, as well as submit requests for additional information. All of NASA's field centers submit information to the TechTracS database as a means of tracking technologies that have potential in the commercial marketplace.

Since their inception in January 1992, the six NASA-funded Regional Technology Transfer Centers (RTTC) have helped U.S. firms investigate and utilize NASA and other federally-funded technologies for commercial and industrial applications. RTTC services include technology sourcing, technology/market analysis and the development of technology transfer and commercialization projects and agreements. Companies seeking new products, improvements to existing products, or solutions to technical problems go to the

RTTCs for help. The RTTCs provide technical, commercial, and general assistance to several thousand customers every year.

Northeast (CT, MA, ME, NH, NJ, NY, RI, VT)

The **Center for Technology Commercialization** (CTC) *<http://www.ctc.org>*. is a non-profit organization based in Westborough, MA. Covering the six New England states plus New York and New Jersey, the CTC currently has seven satellite offices that form strong relationships with the Northeast industry. Operated by the CTC, NASA's Business Outreach Office stimulates business between regional contractors and NASA field centers and prime contractors, with a focus on technologies and facilities available at Goddard Space Flight Center.

Mid-Atlantic (DC, DE, MD, PA, VA, WV)

The **Mid-Atlantic Technology Applications Center** (MTAC) *<http://oracle.mtac.pitt.edu/WWW/MTAC.html>* is located at the University of Pittsburgh in Pennsylvania. MTAC has designed TechScout, a highly specialized set of matchmaking services designed to help companies locate technology/technical expertise within NASA and the federal laboratory system. Close relationships with Goddard Space Flight Center and Langley Research Center allow MTAC to help U.S. firms improve their competitiveness by assisting them in the location, assessment, acquisition and utilization of technologies and scientific engineering expertise.

Southeast (AL, FL, GA, KY, LA, MS, NC, SC, TN)

The **Southern Technology Applications Center** (STAC) *<http://www.state.fl.us/stac>*. is headquartered at the University of Florida in Alachua. Working closely with Marshall Space Flight Center, Kennedy Space Center and Stennis Space Center, STAC helps to spur economic development in each of the nine states in the southeast. To facilitate the transfer of NASA technologies, expertise and facilities, the four NASA entities formed the NASA Southeast Technology Alliance.

Mid-Continent (AR, CO, IA, KS, MO, MT, ND, NE, NM, OK, SD, TX, UT, WY)

The **Mid-Continent Technology Transfer Center** (MCTTC) *<http://www.tedd.org/mcttc>*, under direction of the Technology and Economic Development Division of the Texas Engineering Service, is located in College Station, Texas. The MCTTC, which provides a link between private companies and federal laboratories, reports directly to Johnson Space Center. The assistance focuses on high-tech and manufacturing companies that need to acquire and commercialize new technology.

Mid-West (IL, IN, MI, MN, OH, WI)

The **Great Lakes Industrial Technology Center** (GLITeC), *<http://www.battelle.org/glitec>* managed by the Batelle Memorial Institute, is located in Cleveland, Ohio. GLITeC works with industry primarily within its six-state region to acquire and use NASA technology and expertise, with a special concentration on Lewis Research Center. Each year, over 500 companies work with GLITeC and its affiliates to identify new market and product opportunities. Technology-based problem solving, product planning and development, and technology commercialization services are among the services offered.

Far West (AK, AZ, CA, HI, ID, NV, OR, WA)

The **Far West Regional Technology Transfer Center** (RTTC) *<http://www.usc.edu/dept/engineering/TTC/NASA>* is an Engineering Research Center within the School of Engineering at the University of Southern California in Los Angeles. Utilizing the Remote Information Service to generate information from hundreds of federal databases, the staff then works one-on-one with businesses and entrepreneurs to identify opportunities, expertise and other necessary resources. The Far West RTTC enhances the relationship between NASA and the private sector by offering many unique services, such as the NASA Online Resource Workshop, NASA Tech Opps, and links to funding and conference updates.

The **Research Triangle Institute** (RTI) *<http://www.rti.org>* located in Research Triangle Park, North Carolina, provides a range of technology management services to NASA. RTI performs technology assessments to determine applications and commercial potential of NASA technology, as well as market analysis, and commercialization and partnership development. Working closely with all of NASA's Commercial Technology offices, RTI's efforts have already resulted in the negotiation of 27 licenses to companies and 18 new product introductions based on NASA technologies.

The **MSU-NASA TechLink Center** *<http://www.montana.edu/techlink>*, located at the Montana State University-Bozeman, was established in 1997 to match the technology needs of client companies, with resources found throughout NASA and the federal laboratory system. TechLink focuses on a five-state region which includes Montana, Idaho, Wyoming, South Dakota, and North Dakota. Working closely with public, private and university programs, TechLink provides ongoing support in the process of adapting, integrating and commercializing technology.

NASA currently sponsors four technology commercialization incubation centers across the nation. Programs established at Ames Research Center, Kennedy Space Center, Lewis Research Center, and Stennis Space Center assist start-up companies with commercializing NASA technology. The incubators offer low-cost executive office space and related business services, onsite staff to meet the daily needs of the tenant companies, and access to NASA technology, expertise and facilities.

The **Ames Technology Commercialization Center** (ATCC) *<http://ctoserver.arc.nasa.gov/ATCC/atec.html>* is a physical and virtual small business incubator located in San Jose, California. The ATCC provides opportunities for start-up companies utilizing NASA technologies to grow and become robust high technology businesses.

The **Florida/NASA Business Incubation Center** (FNBIC) *<http://technology.ksc.nasa.gov/FNBIC>* is a joint partnership of Brevard Community College, the Technological Research and Development Authority, and Kennedy Space Center. The mission of FNBIC is to increase the number of successful technology-based small companies originating in, developing in, or that relocated to Brevard County.

The Mississippi Enterprise for Technology is sponsored by NASA, the Mississippi University Consortium and Department of Economic and Community Development, and the private sector. The mission of the Enterprise is to utilize the scientific knowledge and technical expertise at the Stennis Space Center to aid individuals, as well as companies, in applying the technological discoveries of the government to the non-government world.

Managed by Enterpriser Development, Inc., the **Lewis Incubator for Technology** (LIFT) *<http://www.liftinc.org>* is the newest addition to NASA's technology incubators. A strong relationship with staff at Lewis Research Center provides outstanding technology and support resources to businesses in the Cleveland, Ohio area.

The MSU-NASA Techlink Center was established in 1997 at the Montana State University-Bozeman. Diane Cattrell, Techlink Program Specialist, presents a Certificate of Appreciation from the Montana Tradeport Authority to Peter Perna, Techlink Executive Director.

Affiliated Organizations & Services

To complement the specialized centers and programs sponsored by the NASA Commercial Technology Program, affiliated organizations and services have been formed to strengthen NASA's commitment to U.S. business. Private and public sector enterprises build upon NASA's experience in technology transfer in order to assist the channeling of NASA technology into the commercial marketplace.

The road to technology commercialization begins with the basic and applied research results from the work of scientists, engineers, and other technical and management personnel. The NASA **Scientific and Technical Information** (STI) *<http://www.sti.nasa.gov>* Program was established to provide the widest appropriate dissemination of NASA's research results. The STI Program acquires, processes, archives, announces, and disseminates NASA's internal, as well as worldwide, STI.

With the largest collection of aerospace STI in the world, many products and services are available to the aerospace community and public including:

- NASA STI Report Series including Technical, Conference, and Special Publications; Technical Memoranda; Contractor Reports; and Technical Translations;
- STI World Wide Web site with access to the NASA STI database; online ordering of documents;
- Automatic and on-demand document distribution; custom bibliographies; specialized indexing; and
- The NASA STI Help Desk for assistance in accessing STI resources and information.

Free registration with the program is available through the NASA Center for AeroSpace Information (CASI).

For more than three decades, the reporting to industry of any new, commercially-significant technologies developed in the course of NASA's R&D has been accomplished through the publication of *NASA Tech Briefs* *<http://www.nasatech.com>*. First issued in 1963 as single sheet reports and converted to a magazine format in 1975, *NASA Tech Briefs* has been a joint effort of NASA and Associated Business Publications since 1985. The current qualified circulation surpasses 207,000.

The monthly magazine features innovations developed by NASA, industry partners, and contractors that can be applied to develop new or improved products, and solve engineering or manufacturing problems. Authored by the engineers or scientists who performed the original work, the briefs cover a variety of disciplines including computer software, mechanics, and life sciences. Most briefs offer a supplemental Technical Support Package (TSP), available free by downloading from the World Wide Web, which explains the technology in greater detail and provides contact points for questions or licensing

discussions. *NASA Tech Briefs* also contains feature articles on NASA spinoffs, tech transfer resources, news briefs, and application stories.

The NASA **Small Business Innovation Research** (SBIR) *<http://www.sbir.gsfc.nasa.gov/SBIR.html>* Program provides seed money to U.S. small businesses for developing innovative concepts that meet NASA mission requirements. Each year NASA invites small businesses to offer proposals in response to technical topics listed in the annual SBIR Program Solicitation. The NASA field centers negotiate and award the contracts, and monitor the work.

NASA's SBIR Program is implemented in three phases:

- **Phase I** is the opportunity to establish the feasibility and technical merit of a proposed innovation. Selected competitively, NASA Phase I contracts last for six months and must remain under specified monetary limits.
- **Phase II** is the major R&D effort. It continues the most promising of the Phase I projects based on scientific/technical merit, results of Phase I, expected value to NASA, company capability, and commercial potential. Phase II places greater emphasis on the commercial value of the innovation. The contracts are usually for a period of 24 months and again must not exceed specified monetary limits.
- **Phase III** is the process of completing the development of a product to make it commercially available. While the financial resources needed must be obtained outside the funding set aside for the SBIR, NASA may fund Phase III activities for follow-on development or for production of an innovation for its own use.

The SBIR Management Office, located at Goddard Space Flight Center, provides overall management and direction of the SBIR Program.

The **Small Business Technology Transfer** (STTR) *<http://www.sbir.gsfc.nasa.gov/SBIR.html>* Program awards contracts to small businesses for cooperative research and development with a research institution through a uniform, three-phase process. The goal of Congress in establishing the STTR program is to transfer technology developed by universities and federal labs into the marketplace through the entrepreneurship of a small business.

Although modeled after the SBIR Program, STTR is a separate activity and is separately funded. The STTR program differs from the SBIR program in three ways: the funding and technical scope is limited; offerors must be teams of small businesses and research institutions who will conduct joint research; and limitations are set for time and funds in the Phase I and II efforts.

The **Federal Laboratory Consortium** for Technology Transfer (FLC) *<http://www.fedlabs.org>* was organized in 1974. To promote and strengthen technology transfer nationwide, more than 600 major federal laboratories and centers, including NASA, are currently members. The mission of the FLC is twofold:

- To promote and facilitate the rapid movement of federal laboratory research results and technologies into the mainstream of the U.S. economy.
- The approach of the FLC is to use a coordinated program that meets the technology transfer support needs of FLC member laboratories, agencies, and their potential partners in the transfer process.

The **National Robotics Engineering Consortium** (NREC) *<http://www.cronos.rec.ri.cmu.edu>* is a cooperative venture of NASA, the city of Pittsburgh, the state of Pennsylvania, and Carnegie Mellon's Robotics Institute. Its mission is to move NASA-funded robotics technology to industry. Industrial partners join the NREC with the goal of using high technology to gain a greater market share, develop new niche markets, or create entirely new markets within their area of competition.

Scientists and business work together to develop full business plans to guide product development toward the goals of the industrial sponsor. Project teams offer a full range of expertise, from design through marketing. The NREC patents new concepts, provides entrepreneurial training, and streamlines licensing arrangements.

Highlighted below are publications available through NASA's Commercial Technology and STI Programs.

Aerospace Technology Innovation <http://www.nctn.hq. gov/innovation/index.html> is published bimonthly by the NASA Office of Aeronautics and Space Transportation Technology. Regular features include current news and opportunities in technology transfer and commercialization, aerospace technology and development, and innovative research.

NASA Spinoff <http://www.sti.nasa.gov/tto/spinoff. html> is an annual publication featuring current R&D efforts, the NASA Commercial Technology Program and successful commercial and industrial applications of NASA technology. In addition to the print version, the publication maintains a World Wide Web site which includes online versions of the publication and a searchable database of products and services derived from NASA technology that have been featured since 1976.

NASA's STI Program generates, among others, these important publications:

- The *Scientific and Technical Aerospace Reports* (STAR) is an electronic abstract journal, published biweekly, that announces documents recently entered into the STI database.
- The *NASA Thesaurus* contains the authorized subject terms by which the documents in the STI databases are indexed and received.
- The *NASA Patent Abstracts Bibliography* is a semiannual publication containing comprehensive abstracts of NASA-owned inventions covered by U.S. patents and applications for patent.
- The *STI Bulletin*, now an online-only publication, is the official newsletter of the NASA STI Program.

The NASA Center for AeroSpace Information in Hanover, Maryland is home to the Scientific and Technical Information Program as well as the NASA Spinoff Team. Pictured left to right are members of the Technology Transfer, Publications and Graphics Team: Danielle Israel, Editor, Doris Wahl, Publications coordinator, Walter Heiland, Spinoff Project Manager, Deborah Drumheller, Senior Publications Specialist, Amy Harding, Senior Editor, and John Jones, Senior Graphic Designer.

Commercial Benefits—Spinoffs

During the past four decades, a wide-ranging array of Research and Development (R&D) activities has been conducted by the National Aeronautics and Space Administration. They continue to provide large volumes of Scientific and Technical Information; and each year, hundreds of transferrable new technologies have resulted from such efforts.

The taxpayers investment has been immense; the benefits returned to mankind even larger. Although NASA's budget has decreased, the versatility in R&D efforts continues to spring forth sophisticated technologies ready for application in the development of commercial products. Thus, industry profits from the agency's Commercial Technology Program with its array of services geared to assist in the maintaining global competitiveness.

The following pages demonstrate once again the U.S. engineering ingenuity assisted by NASA's Commercial Technology Network. Extensive outreach activities and bold partnerships between application and development engineers and NASA have created many new and diverse commercial products and services—we refer to them as "spinoffs," and some of them are highlighted here.

BEATING BACK BACTERIA

Open those pearly whites...wider, please. Going to the dentist is always a less-than-welcomed experience. But a new bacteria-beating dental unit means that the end result of cleaner teeth will be more satisfying.

MRLB International Incorporated, River Falls, Wisconsin, has designed DentaPure®, a dental waterline purification cartridge that relies on water purification technology developed for NASA. MRLB's unit can clean and decontaminate water as a link between filter and dental instrument. The purification cartridge can be installed in seconds and changed, not daily, but once a week. For use on high-speed dental tools and other instruments, the cartridge is easily installed on all modern dental unit water lines.

As an answer to contaminated dental unit water, the product furnishes disinfected water, maintaining water purity even with suckback. Complete with a tiny membrane, the cartridge is crafted to remove or destroy bacteria to levels that meet or exceed American Dental Association recommendations for dental unit water quality.

The cartridge for dental use incorporates a resin technology developed by Umpqua Research of Myrtle Creek, Oregon. Umpqua has been awarded a number of Small Business Innovation Research (SBIR) contracts by the Johnson Space Center, aimed at providing air and

water purification technologies for human-carrying missions in space. Also, bacterial filters fabricated by Umpqua are used to ready the life support backpacks utilized by NASA space-walking astronauts. Umpqua's research has been a plus for NASA, with the company providing the only space certified and approved-for-flight water purification system. That system has flown on all Shuttle missions since 1990.

In 1993, the company demonstrated a regenerable purification unit, with NASA granting an exclusive license to the company, permitting the patented product to be adapted for commercial sale. Commercial applications of the SBIR-supported technology are promising—in schools, hospitals, and in countries around the globe. This same technology is scalable for use in municipal water treatment plants, yet portable in design to help far-flung villages have safe, drinkable water. Furthermore, the system can find utility in emergencies, like floods and other natural disasters, when a city's infrastructure has collapsed.

An Umpqua innovation is the microbial check valve (MCV®). The device prevents back contamination of a drinking water supply by microorganisms, thanks to a flow-through cartridge containing iodinated ion exchange resin. It was found that, in addition to the microbial contact kill, the resin imparts a biocidal residual elemental iodine concentration to the water. That concentrate is

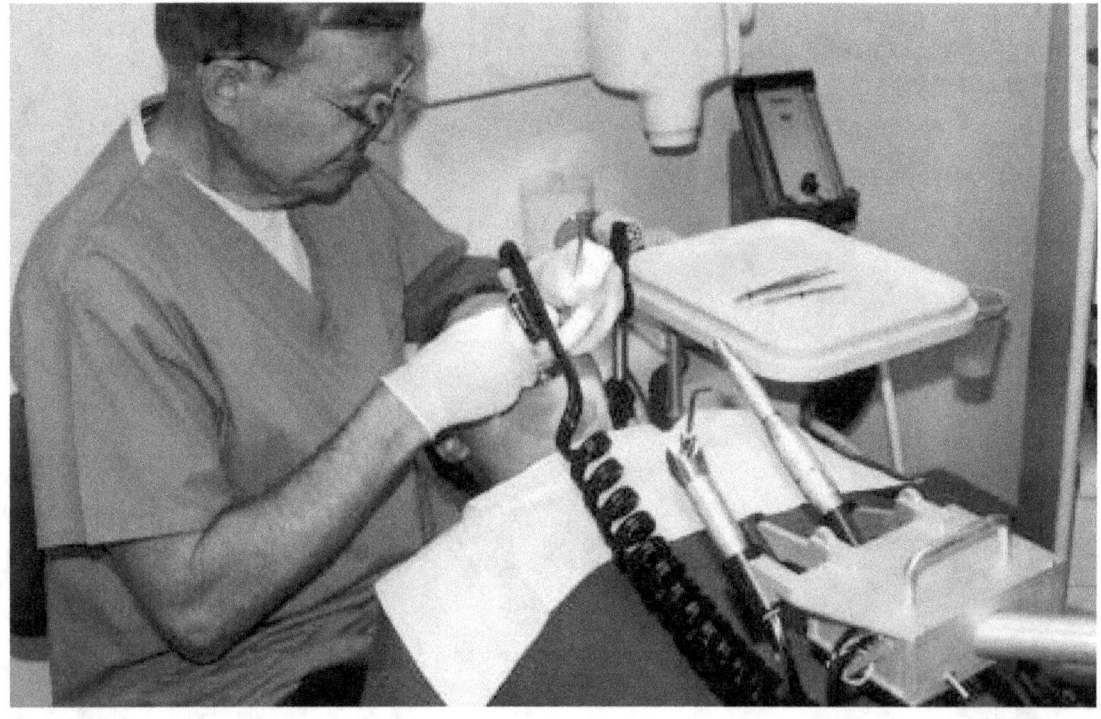

A visit to the dentist gives this patient the cleanest of water while dental tools do their work. NASA technology for purifying water has been applied to dental care.

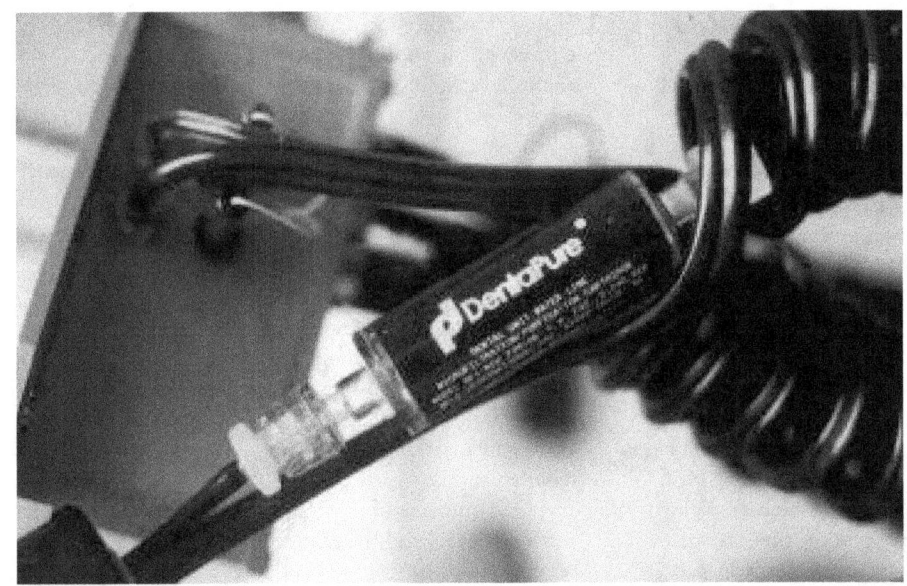

Microbial contamination in dental water lines has been widely reported for many years. The DentaPure® cartridge is the answer to eliminating contaminated water.

sufficient to maintain a quality of drinkable water. Tests have shown that a single resin can be regenerated a hundred or more times.

Umpqua's valve and resin has been adopted by NASA as the preferred means of drinking water disinfection aboard U.S. spacecraft. Canisters designed by the company are not only used on Space Shuttle missions, but for ground-based testing of closed life support technology, and are being deployed aboard the International Space Station. Iodine was selected by NASA as the disinfectant of choice because of its lower vapor pressure and reduced propensity for formation of disinfection by-products in comparison to chlorine or bromine.

Private sector commercialization of Umpqua's technology developed under NASA contract has been impressive. According to World Health Organization estimates, more than one billion people lack daily access to safe drinking water. Thirty-five percent of all deaths in developing countries are directly related to contaminated water. Chlorination is a common method of disinfecting drinking water. But that means a need for electricity, pumps, tanks, and technical expertise, which often are unavailable in rural or remote locations in emerging nations.

One potential solution is the simple flow-through canister approach. Umpqua technology is ideal for societies that lack the infrastructure necessary for chlorination. No energy is required. The cartridge may simply be installed in line with the drinking water flow. Residual elemental iodine is imparted to the water on demand. Another plus is that the valve and resin can perform over a wide range of temperatures and flow rates.

"It's particularly useful out in the field where you can't have somebody there to monitor microbial control technology and you need something simple and rugged enough for a variety of environments," says co-inventor Clifford Jolly, who directed the fundamental development of the regenerable resins at Umpqua.

Vector Ventures, Inc. of Nevada is developing the first large-scale application of Umpqua's Space Shuttle-derived technology in Vietnam. Under a Memorandum of Agreement with that country, Vector will install water purification equipment to support 50-70 million people. Umpqua is also negotiating with other nations, such as Indonesia, India, and China to further expand the reach of its purification technology.

®DentaPure is a registered trademark of MRLB International.

®MCV is a registered trademark of Umpqua's Research.

CANCER DETECTION DEVICE

Space medical specialists have long been intrigued by how and why microgravity affects the immune system. Preliminary evidence from Space Shuttle flights suggests that immunity is depressed. To help decipher this medical mystery, NASA sought hardware development of a machine that could separate and examine cells rapidly. Space engineers quickly ran into a snag. The device to do the work, a flow cytometer, was far too big and cumbersome—about the size of a pool table—to loft into orbit on a space station.

NASA, with the Kennedy Space Center as the lead center, and the American Cancer Society (ACS) teamed in 1989 to develop a far more compact flow cytometer.

A Memorandum of Understanding was signed, establishing a means by which space technology could be formally transferred to the medical profession. The NASA Technology Applications team at the Research Triangle Institute supported this partnership in the first systematic approach to matching a list of medical needs with space technology.

NASA researchers were delighted with the joint research endeavor to develop an advanced flow cytometry instrument. This kind of device could support biomedical experiments aboard the space station while advancing medical knowledge in cancer detection and treatment here on Earth.

Flow cytometry, the initial project undertaken by the partnership, is a process in which cells in suspension flow through a sensing region where light signals—indicating important biologic properties—are generated and evaluated by photodetectors. The cancer-fighting benefits of flow cytometry include the ability to evaluate cancer cells very early and to determine several important features, including the sensitivity of those cells to different chemotherapy drugs, the ability of the cells to grow, and their capacity for spread. Better and more timely strategies in the fight against cancer was a main objective of the research.

The challenges in developing multichannel flow cytometry were addressed in a NASA/ACS workshop. Technical improvements needed were improved signal processing for multichannel analysis of optical emission spectra; reductions in complexity, size, power requirements; and numbers of optical sensors, simplification of sample preparation, and expert system software. The University of Miami was instrumental in defining requirements and early testing of prototypes.

One outcome of the long-standing cooperation between NASA and the ACS led to a new tool—a high-resolution flow instrument designed specifically for DNA analysis of solid human tumors.

RATCOM, Inc., of Miami, Florida, started offering the DNAnalyzer®, the first commercial instrument stemming from the NASA/ACS partnership, the Space Station In-Flight Cytometry Project. RATCOM pioneered a new triangular flow cell technology. This work bestows the DNAnalyzer with twice the resolution and three times more uniformity on a day-to-day basis than results on the same samples from older flow cytometer technology. Testing of the new instrument in 1997 fully confirmed the advancement in flow channel design.

"There has been a major improvement in resolution and reproducibility," says Richard A. Thomas, President of RATCOM. "This translates into a better understanding of the nature of the patient's tumor, and therefore better treatment. The improvement is not marginal. The instrument allowed positive confirmation of tumors, which were only suspicious by the older technology in twenty percent of the cases in the 170-patient study," he says.

Advanced flow cytometry has the potential to become a significant tool in fighting cancer. Other potential uses involve the clearing of cancer cells from bone marrow prior to retransplantation of the patient's own bone marrow after chemotherapy. Patients who have leukemia or AIDS may also be followed by means of flow cytometry to determine the effectiveness of treatment.

Thomas gives credit to the cooperative project with NASA, enabling him to commercialize a flow cytometer for cancer diagnosis.

®DNAnalyzer is a registered trademark of RATCOM, Inc..

NASA cytometer project for the Space Station spurred the development of this instrument—important for cancer diagnosis—that can properly classify tumors.

SIT UP STRAIGHT

Laying down on the job is every astronaut's starting position en route to space.

Reclining crew members are on their backs for several hours awaiting launch. Bad posture or protracted activities during work, even for a space traveler, can cause strain and fatigue.

One would think if a person's work was confined to the microgravity of space, such a free-floating experience would cause little worry. But astronauts in training for the Space Shuttle and space station programs receive the do's and don'ts of correctly positioning their bodies.

A posture video analysis tool (PVAT) was developed at the Johnson Space Center (JSC). The video uses scenes from Space Shuttle flights to classify limiting posture and other human factors in the workplace. NASA needed a low-cost, reliable method of collecting data on astronaut postures from non-scientific mission video. The traditional "paper and pencil" video analysis methods are very subjective. The current state-of-the-art digitizing system requires a predefined view from spacecraft cameras as well as specific reference points to classify working posture.

With the new video tool, researchers use regular non-scientific shuttle videos to gather precise information about astronaut working postures and movements. The software side of PVAT provides data that ranks postures for certain tasks and duration of time in a position.

Button-driven software and an interactive menu collect information on a variety of postural parameters— body orientation, body part movement, severe or mild flexation rating, and task description. Once all the entries are made, analysis begins with the touch of a button. The tool also includes a terminology library. Selected posture classifications are illustrated in animated form. Data reduction summaries and report capabilities are incorporated to round out the video software product.

BioMetric Systems of Houston, Texas, has received an exclusive license from JSC to further develop PVAT for use by non-aerospace industries, such as hospitals, physical rehabilitation facilities, insurance companies, sports medicine clinics, oil companies, manufacturers and the military. BioMetric learned of the software from the Mid-Continent Technology Transfer Center (MCTTC), one of NASA's Regional Technology Transfer Centers.

When used with videotape, the PVAT system allows BioMetric to perform ergonomic analyses of people in the workplace. Examples of how the system can be used include gathering information to determine the correct height of a keyboard or the proper height of a chair or stool. The system also will help a company determine if employee ailments are caused by inefficient equipment setup or poor worker technique.

"PVAT is unique because it provides a fast and simple way to collect and classify working postures, even from videos not recorded specifically for experimental analysis,"

says BioMetric Systems President, Candace Caminati. "We are excited about PVAT's human factors design and analysis potential in a variety of commercial industries," Caminati says.

The human factors engineering company is augmenting the JSC PVAT software systems for use on different types of computer operating systems. Assistance in this upgrade was provided by JSC's Technology Transfer and Commercialization Office.

After obtaining the exclusive license from NASA, the company embarked on a technology research improvement program with New Mexico Highlands University. As part of BioMetric's product development program, the company submitted a formal proposal to NASA's Small Business Technology Transfer (STTR) program for cooperative research and development with a research institution.

BioMetric Systems has grown into a respected manufacturer and distributor of computer software for scientific and engineering analysis of human factors relating to medical, petrochemical and space operations/simulations applications. The company provides training, workshops, and consulting on ergonomic design, soft tissue injuries, sports medicine, athletics, personal injury, workman's compensation, exercise physiology, stress analysis, and return-to-work programs.

As a woman-owned small business, BioMetric was the first Native American company to license commercial technology with JSC.

BioMetric has settled on three versions of the PVAT software service package. One version provides instructions for clients to setup, use, and analyze ergonomic data themselves; another lets clients use the package and return the results to BioMetric for analysis; and a third version is designed to have BioMetric perform the work onsite, from setup to analysis.

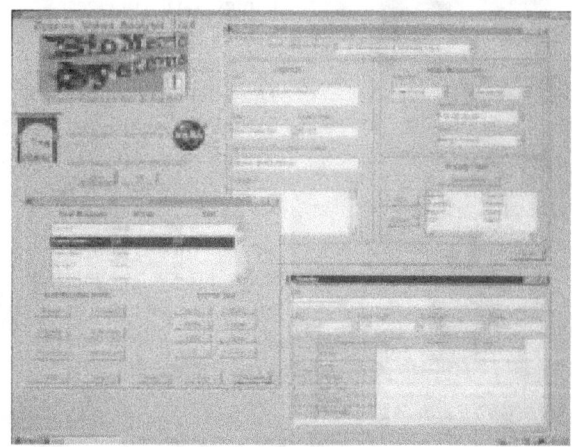

The PVAT system helps companies perform ergonomic analyses of employees in the workplace. This screen shows extensive activity type and body movement criteria that can be applied to the subject being evaluated. The system is based on a video analysis tool for collecting data astronaut postures.

COMPUTER-BASED MEDICAL SYSTEM

It would be a major spaceborne dilemma. Halfway to the planet Mars, a crew member complains of an acute, internal pain. Accurate diagnosis of this potentially life-threatening condition is required quickly. What can be done? The 21st century answer to the problem is now available.

A medical information and diagnostic support system designed for space travel has found new earthbound applications. Supported by contracts through the Kennedy Space Center, early work on the medical communications system was done under the NASA project, System 2000, also known as S2000.

Originally prompted by the need to help astronauts diagnose their own illnesses, S2000 has led to new health care hardware and software—an electronic medical library and record keeping system. Doctors can now better inform patients about their care, keep paperless medical records, prevent dangerous drug interactions and confirm diagnoses.

Over a period of 15 years, under NASA funding, Ralph Grams, MD, a medical systems specialist at the University of Florida's College of Medicine, developed the computer hardware and software system. To stimulate commercial adoption of the concept, NASA relinquished their copyright interests in the software development to the University of Florida.

In 1995, Grams founded SyMed, Inc. of Gainesville, Florida, a company which would dramatically improve the effectiveness of the system for single/multiple physician practices and clinics. That same year, SyMed finalized a licensing agreement with the University of Florida to obtain the exclusive, worldwide rights to the intellectual property for a 25-year period. The agreement also grants SyMed the sole ownership of any improvements, enhancements, or modifi-

The physician inspects a mole on patients arm with a novel computerized microcomputer. A system that came out of a NASA medical communications project.

cations made to the S2000 software during this same time period.

"Computer-based, medical interactive care systems will revolutionize how we take care of patients," says Grams. "The physical examination room will be transformed into a multimedia teaching studio, where physicians can pack into a brief exam a great deal of useful medical information that patients and their families can leave with," he adds.

At the very heart of SyMed's S2000 is an advanced "patient friendly" feature, microcomputer access to the latest medical information from twelve reference books, some 1,000 medical journals, and an atlas of human anatomy. The program features digital dictation and transcription, automated storage and retrieval of patients' records, information on more than 20,000 drugs and their side effects, and video pictures and sound bytes recorded during the physical examination for the patient's medical chart.

"Patients will see their health problems and therapy explained in full color animation and surround-sound, with their lab and radiology information retrieved instantly and projected on a large-screen TV or wall screen. Patients will review this information alongside their doctor, so they can become thoroughly informed about their health problems and how to best treat them," Grams explains.

The technology spinoff company reached out to the Southeast Regional Technology Transfer Center, operated by the Southern Technology Applications Center (STAC) at the University of Florida. STAC staff helped SyMed to establish business affiliations with an international sales network. STAC also provided a technology assessment of SyMed's S2000 and reviewed their business plan, and brought in a management consultant to focus on raising capital and finding a strategic partner. STAC introduced Grams to Ernest Moyer, a local businessman and technology commercialization specialist. Moyer's talents enabled the finalizing of a fair and equitable license agreement with the University, raised the necessary initial capital required by the company to proceed, and took on administrative and financial tasks that an early-stage company needs to perform.

In April 1997, SyMed announced it had negotiated a strategic alliance with Soft Computer Consultants of Palm Harbor, Florida. This alliance is expected to strengthen and open up SyMed's S2000 capabilities for reaching out to hospitals and other large health care institutions.

"An automatic, interactive medical office system would let physicians make better use of their time during the brief doctor/patient encounter and actually help them build stronger relationships and trust with their patients," Grams says.

CHROMOSOME ANALYSIS

"Probing the future of genetics," is the thrust behind Perceptive Scientific Instruments, Inc. (PSI) of League City, Texas.

The company's initial foray into the field of genetics was sparked by work taking place at two different NASA centers. At the Jet Propulsion Laboratory, technicians were handling image processing of space probe pictures sent back to Earth. Meanwhile, computerized cell analysis studies were underway at the Johnson Space Center. This blend of interests served as the nexus for PSI, with many of the initial company employees being former NASA image processing specialists. Digital imaging techniques have been developed and refined for use in a variety of medical applications, including the diagnosis of disease.

Today, PSI is considered one of the largest international companies in the world dedicated solely to the supply of digital imaging systems for automated cytogenetic, karyotyping, and molecular genetic (DNA probe) applications.

Cytogenetics is the science that deals with the relation of humans cells—and their building blocks—to heredity. Karyotyping is a process employed in analysis and classification of chromosomes, the bodies within a cell that carry the genes which determine heredity.

In October 1997, PSI unveiled its latest addition to the PowerGene™ line of automated chromosome analyzers. The announcement came at the American Society of Human Genetics Conference, the largest gathering of genetic experts of its kind.

The new equipment takes advantage of recent advancements in multiplex fluorescence-in-situ hybridization genetic analysis methods. The fully integrated system is capable of 24-color karyotyping. Utilizing highly specific fluorescent reagents and a set of finely tuned optical filters, images of chromosomes during the metaphase of mitosis are captured, processed, and karyotyped. Each chromosome pair, including the sex chromosomes, is displayed in its own unique color.

"This exciting new development has the potential to change the way in which most cytogenetics studies are performed today. It should be particularly useful in the field of cancer genetics, where this new approach can radically simplify the detection of chromosome rearrangements."

The automatic system guides the user intuitively through the entire karyotyping process. Unique image enhancement facilities produce an on-screen and hardcopy image, yielding the highest quality chromosome band visualization possible. Editing tools give the user an ability to zoom in for easy and accurate separation of chromosomes. Confirmation of overlap separation of each chromosome is visualized by colored chromosome

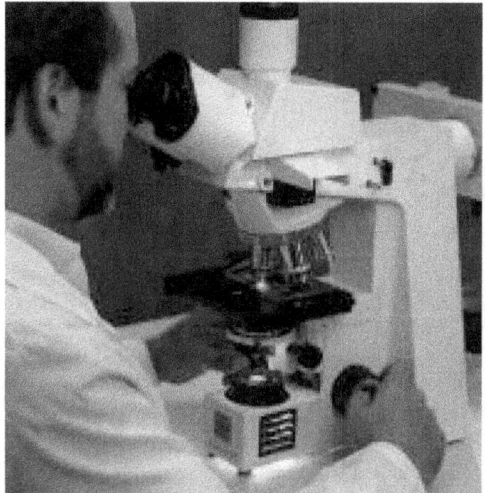

The PowerGene™ system is the latest addition to Perceptive Scientific Instruments' automated chromosome analyzers. NASA image processing techniques provide the foundation for the latest of chromosome analysis and genetic instrumentation.

boundaries. A review window feature permits up to three karyotypes, from the same or different cases, to be inspected. Each image can be scrolled simultaneously row-by-row.

While powerful in its ability to offer complex digital imaging through sophisticated pattern recognition techniques, the system operates on the friendliest of user-friendly interfaces. Compare the past to today. Formerly, karyotyping was a laborious, time-consuming task that involved photographing the chromosomes through a microscope, then manually cutting and pasting the images to put together a classification. A job that once demanded hours, can now be completed in less than ten minutes. That increases productivity, at the same time lowering costs in the research laboratory.

PSI started a collaboration with top scientists to investigate new automation techniques designed to assist in the detection of genetic abnormalities. Two projects are aimed at monitoring the progression and remission of leukemia, and developing new techniques for the detection of commonly occurring aneuploidies in newborns.

PSI has already installed genetic imaging systems in more than 400 laboratories in over thirty-six countries worldwide. The company has automated five of the largest commercial genetic testing laboratories in North America with over 140 networked karyotyping workstations.

PSI also installed what is believed to be the largest probe and karyotyping system network in Asia—ten full analysis workstations. The company has sold its product line in more than thirty-five countries worldwide and has installations in more than 400 user sites.

™PowerGene is a trade mark of Perceptive Scientific Instruments.

MEDICAL PRACTICE MAKES PERFECT

Healthcare providers strive to provide more efficient service in a competitive and cost-conscious world. The pressures of managed care can put a tremendous strain on doctors, staff, and certainly the patients themselves. Paperwork can overwhelm all concerned, just in monitoring treatment effectiveness and reimbursement.

Cedaron Medical, Inc., Davis, California, manufactures a range of products to increase clinical productivity, thereby enhancing patient care.

Cedaron was founded in 1990 on the award of a Small Business Innovation Research (SBIR) contract from Johnson Space Center to provide a hand testing and exercise unit for use in space. From that research, Cedaron struck out as a manufacturer of orthopedic therapy devices. But the business saw on the horizon an emerging and promising market in outcomes analysis.

Outcomes analysis measures a hospital's performance in several ways:

- Patient response to the care provided;
- Measuring costs and average length of stay against a comparable treatment;
- Gauging such things as strength and range of motion in orthopedic and other treatments.

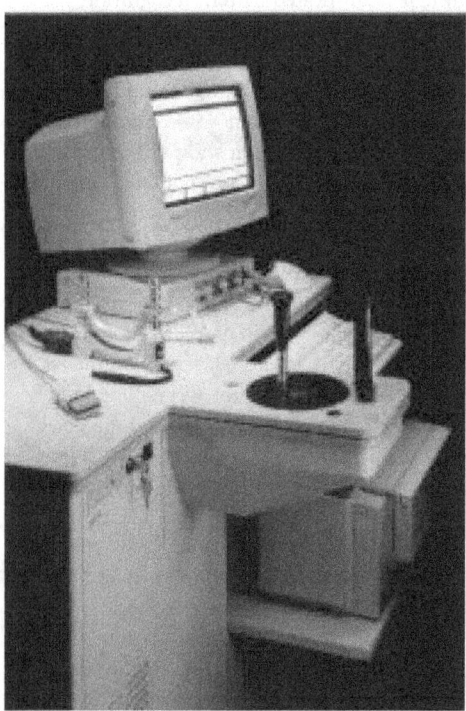

Cedaron Medical's Dexter Outcomes workstation provides a friendly data collection system for the occupational therapy, physical therapy, orthopedic surgery, and plastic surgery fields. The system is based on a Small Business Innovation Research contract to develop a system to monitor upper extremity function of astronauts during space flight.

A few years ago, Cedaron expanded its medical performance work by creating computer software for the "paperless office" of today. This specially-devised software is aimed at measuring the effectiveness of medical treatments in occupational therapy, physical therapy, orthopedic surgery, and plastic surgery. Similar applications are also being developed for the field of cardiology. Above all, Cedaron's Dexter Outcomes products are crafted to minimize paperwork shuffle, and to establish a more cost-effective and efficient medical care facility.

One Cedaron product is a specially-devised physician's workstation. Not only can it manage patient records more effectively, the workstation speeds up communications between referring physicians and therapists. Keeping track of progress reports, updating patient records, and recording daily notes are among the workstation's many features.

For the patient, Cedaron has also designed an outcomes workstation. This computerized, patient-friendly system is replete with a touch-screen interface to collect outcomes data quickly and simply from a patient. Questions are asked about medical history, functional status, symptom severity, general health, and satisfaction with healthcare delivery. The responses are stored in an outcomes database that can be linked to other computers.

Other products from Cedaron include an upper extremity tool kit, a tabletop workstation for extremity and spine evaluation; and a hand center to conduct exams, strength tests, and impairment rating protocols—automatically documenting the results.

In December 1997, Cedaron announced that it had been included on the Joint Commission on the Accreditation of Healthcare Organization's list of acceptable performance measurement systems for hospital accreditation. As part of new accreditation guidelines, hospitals must enroll in, and have a certain percentage of their patients included in, programs that measure the outcomes of healthcare delivered by the hospital.

Springboarding from its early work with NASA, today Cedaron has computer systems installed in Asia, Europe, South America, and across the United States.

"Since our inception, Cedaron has focused upon intuitive, interactive patient evaluation and medical documentation systems," explains Karen Bond, President and CEO of Cedaron Medical. "Our systems automate the process of healthcare outcomes without adding to the workload of the staff. And, our customers tell us that patients enjoy interacting with our systems. It really is a win-win technology for both hospitals and patients," she says.

A STIRLING IDEA

Super cool! Low temperature refrigerators, medical diagnostic equipment, and sophisticated electronics—all these are benefiting from cryocooling technology.

Stirling Technology Company (STC), Kennewick, Washington, designed a line of cryocoolers under Small Business Innovation Research (SBIR) contracts with Goddard Space Flight Center (GSFC) and Marshall Space Flight Center (MSFC).

Stirling engine technology is rooted in the creative tinkering of Scottish brothers, Robert and James Stirling. They invented the Stirling engine in 1816. STC has advanced Stirling engine technology, bringing it from the 1800s into the Space Age. These type engines are unparalleled in energy efficiency. They approach the limits set by the laws of thermodynamics more closely than any other power system. Stirling engines need less heat energy to generate a given power of output. Versatile in performance, Stirling engines can be reversed to make refrigerators, cryocoolers, or heat pumps. So efficient is the engine, it can chill to cryogenic temperatures.

GSFC awarded the SBIR funds to characterize spacecraft sensor/instrument cooling hardware, predicated on the high performance of a liquid helium temperature refrigeration system. STC culled together company talent in thermodynamic, dynamic, and mechanical design, producing for GSFC a low-temperature approach.

MSFC's needs were different. SBIR work for MSFC was geared toward a spacecraft food freezer/refrigerator. NASA requirements centered on a refrigerating unit that ran at a cooling potential or average freezer heat load of 115 thermal watts, and operated at a chilly temperature of -15 degrees Fahrenheit. An upgrade of the modular refrigeration unit is slated to be aboard the International Space Station. The innovative Stirling cycle refrigerator approach gave NASA a high performance, rugged, and quiet refrigerator system. Moreover, the system does not require ozone-depleting chlorofluorcarbons, better known as CFCs.

A new linear motor, invented by Dr. Syed Nasar from the University of Kentucky, proved key. The motor accommodated low-cost mass production assembly and fabrication techniques. STC tests and refinement of this motor design proved crucial in the commercialization of the cryocoolers.

The result? The BeCOOL™ line of low temperature refrigeration equipment that sports long life, low maintenance, high reliability, and attains high safety characteristics. All of these features combined to make the newest cryocooler hardware attractive for a variety of commercial applications, such as controlling computer temperature and for laboratory experiments. These attributes of STC's BeCOOL cryocoolers have a heritage based in the tough, technology-pushing prerequisites of NASA.

STC has been able to produce a high-capacity linear drive cooler, one with features that satisfy commercial demands. It has a demonstrated operation life exceeding 30,000 maintenance-free hours. The product will be initially sold to laboratories that require cryogenic refrigeration and for medical applications. The low noise level of the cooler permits its operation in work areas.

STC believes that niche markets are likely to evolve for power generators that are highly-efficient, reliable, maintenance-free, multifuel compatible and produce ultra-low emissions. Supported by company research funds, 10-watt and 350-watt power generators have been built. Multiple units have been sold to government and commercial customers for evaluation purposes. STC's forecast is a demand for turn-of-the-century generators that offer a capacity in the 3-kilowatt range.

Since incorporation in 1985, STC has received over $22 million in research and development contracts from both government and commercial clients. Successful completion of these contracts has enabled STC to design and build unique products, earning the company the reputation for providing long-life, maintenance-free, high-efficiency Stirling machines.

™BeCOOL is a trademark of Stirling Technology Company.

Stirling Technology Company engineer readies the cryocooler. NASA requirements called for low-temperature equipment to run sensors to achieve refrigeration levels for a space-rated freezer.

ON THE WING: A BUSINESS-CLASS JET

It is the fastest, most efficient business jet ever to race across the sky. Capable of flying at near supersonic speeds, an exciting addition to Cessna's fleet of private aircraft has pushed the envelope of general aviation to new performance levels.

NASA's Langley Research Center and the Cessna Aircraft Company of Wichita, Kansas have had a long, beneficial relationship. Starting with use of supercritical wing technology developed by NASA in the 1970s, Cessna has repeatedly called upon NASA's know-how and use of its test facilities. On this latest venture into business-class jet aircraft building, a highly swept, second-generation supercritical wing was designed, capable of Mach .92 and supporting the aircraft to a maximum altitude of 51,000 feet.

To help in the development of the unprecedented aircraft, Citation X, Cessna retained as a special consult-ant, Richard Whitcomb, a retiree from NASA Langley and the inventor of the supercritical airfoil.

A supercritical airfoil enhances an airplane's performance in a critical flight area—the transonic regime (speeds near Mach 1). High-speed subsonic aircraft can experience mixed subsonic and supersonic airflow, and at some point over the wing, airflow exceeds the speed of sound. At this point, the airflow surface suddenly changes, creating a standing shock wave. In turn, this phenomenon results in excessive drag, and therefore a loss in efficiency. Supercritical wings change the shape of the airflow by flattening the upper surface of the wing, which minimizes the effect of the shock wave. Low drag contributes to an aircraft's fuel efficiency.

NASA joined forces with Cessna to assist in the development of the largest, most complex aircraft ever pursued by the company. As a U.S. airframe manufac-

New Cessna business plane benefited by NASA wind tunnel research, work in composite materials, and aerodynamic expertise.

turer, Cessna had full access to the research, personnel, and facilities of NASA. For instance, Langley's one-of-a-kind Transonic Dynamics Wind Tunnel proved extremely valuable. The tunnel has a 16-foot square test section and is capable of operating at Mach numbers up to 1.2. Along with gathering data on the private jet's wing design, the facility also helped in measuring the unsteady aerodynamic pressure from air loads on the new business jet's wing design. The wind tunnel data were used to validate Cessna's flutter analysis.

Flutter is defined as an unstable, self-generated oscillation of an airfoil and associated structure. If left unchecked, flutter can practically shake an aircraft apart. While conventional, lower-speed business jets test flutter boundaries in flight test, more intensive testing was considered prudent for the new and unique Cessna aircraft prior to flight testing.

Winging its way across a sky near you, Cessna's newest addition to private business jets tapped NASA aeronautical research and technology development.

A quarter-scale fuselage and wing model underwent weeks of exhaustive flutter and unsteady pressure tests in the NASA tunnel. Langley's wind tunnel was perfect for the job. According to Cessna's Engineering Director, Ellis Brady, no other general aviation test model had ever measured as much of this type of data. "These tests gave us an added measure of assurance and confidence in the safety of this aircraft," Brady notes.

NASA also collaborated on the Cessna project by providing computational time on its number-crunching Cray computer. This aspect of testing was particularly useful to Langley engineers because the business jet's high-speed wing provided invaluable correlative test data for NASA computer programs used for evaluating transonic structural dynamics.

Langley also aided Cessna in the acoustics area. Research conducted at Langley, and at Virginia Polytechnic Institute and State University under a NASA grant, contributed to the design and implementation of the business jet's active noise control system. This approach reduces or nulls out noise by introducing a countering soundwave at the same level and frequency, but shifted out of phase.

Much less directly, Langley also contributed to the aircraft's lightweight structure. Advanced composite material technology applied to the Cessna craft can be traced to pioneering Langley research in developing fiber-resin material systems. Over two decades of study in this area have yielded a dependable database, at the same time establishing that composite materials, versus aluminum, can be safely used on aircraft.

The collaboration between Cessna and NASA was award-winning in many ways. The aircraft was recognized as the top aeronautical achievement in the United States for 1996, with Cessna receiving a prestigious Robert J. Collier Trophy.

As the fastest commercially-built aircraft in the United States, only one non-military aircraft presently in service worldwide is faster—the Concorde. The business plane combines cruising speed with true intercontinental and transatlantic range. Taking just 67 months from concept to design completion, the Cessna Citation X aircraft has already accumulated several thousand flying hours, and is winging its way into the record books.

GETTING TO THE POINT IN PINPOINT LANDING

A high-performance navigation system used primarily for automatic aircraft touchdowns promises centimeter-level landing accuracy.

Founded by alumni of the Stanford University Department of Aeronautics and Astronautics, IntegriNautics of Palo Alto, California has commercialized a precision landing system. The work has been assisted by Langley Research Center's Small Business Technology Transfer (STTR) Program.

But the real genesis of the idea was fostered by Stanford University work on a satellite test of Einstein's General Theory of Relativity. Called Gravity Probe B, this soon-to-orbit NASA spacecraft will rely on the Global Positioning Satellite (GPS) system for both precise orbit location and spacecraft attitude determination. To do so, researchers at Stanford designed new high-performance attitude-determining hardware that used GPS signals, then flight tested the system on both spacecraft and aircraft.

It was this space project that sparked new thinking on a precision touchdown concept called the Integrity Beacon Landing System.

During a four-day period in October 1994, the idea was put to the test on Runway 35 at NASA's Crows Landing Flight Facility in California. These tests proved the validity of using what IntegriNautics terms integrity beacon "pseudolites." Compact in size, the ground-based low-power transmitters each fit entirely on a circuit board the size of a credit card. Capable of running on a 9-volt battery for over 12 hours, the inexpensive devices transmit just a few microwatts of power, emulating a GPS satellite. The beacons were situated in pairs on either side of the approach path to the runway. Power of the broadcast signals from the pseudolites was set low, measurable only inside a "bubble" emanating from the transmitter.

Using signals from orbiting GPS satellites and the ground-generated pseudolite signals, 110 autopilot-in-the-loop landings of a United Airlines Boeing 737 were completed. The integrity beacons provided consistent accuracies on the order of a few centimeters during each of the autopiloted runway touchdowns. The successful series was sponsored by the Federal Aviation Administration (FAA) as part of that agency's satellite navigation program.

Evaluation of test results provides confidence that the level of integrity yielded by satellite positioning and the ground-based monitors would improve passenger safety. High integrity of the beacon landing system translates to just one failure in a billion approaches.

The company envisions a market for precision navigation, based on levels of performance beyond those provided by the current GPS satellite system. GPS is a worldwide navigation system providing 100-meter accuracy in raw form.

IntegriNautics is now developing technology and products for a range of FAA, NASA, and Defense Department requirements, as well as commercial and international customers. Indeed, the first commercial sale of pseudolites took place in July 1997.

The first pseudolites were sold to customers for use in applications such as aircraft landing research, indoor GPS-based sensing, and robotic vehicle control. A key element of the IntegriNautics product selection is a processor that acts as a turnkey, high-performance positioning system. Several types of GPS pseudolites are presently offered. Work is continuing with Stanford through the STTR program, run through the Langley Research Center.

Built to operate indoors and outdoors, even in inclement weather, precision navigation devices are foreseen by IntegriNautics to have countless applications. Just a few potential pseudolite uses are obvious, such as commercial and general aviation aircraft, agricultural vehicles, open-pit mining, and automobiles.

Boeing 737 conducts series of automatic aircraft touchdowns relying on a new precision navigation system.

FOAM: THE "RIGHT STUFF" FOR EXTREME ENVIRONMENTS

Minimizing fire hazards on the Space Shuttle has inspired production of a polyimide foam that is finding secondary commercial benefit as a lightweight thermal and acoustic insulating material.

Early in NASA's history, Johnson Space Center engineers called for a decrease in the flammability, smoke and toxicity of materials used in spacecraft. That need was adopted in the construction of the Space Shuttle fleet. Initial Space Shuttle uses of the foam were for thermal insulation, as well as packaging and protecting fragile equipment.

Manufactured for use by NASA, Inspec Foams Inc. (formerly named ImiTech) supplied the lightweight, fire-resistant Solimide® polyimide foam. Inspec Foams is headquartered in the Dallas, Texas area. Two big pluses offered by the product are its exceptionally low density and low flammability.

There is no question, the harshness of space is tough on any material. Yet Solimide foam remains flexible and resilient, despite the wild temperature swings of the space environment. It has become the ideal choice for lightweight insulation in space applications. One such use is cryogenic insulation for fuel tanks on major rocket propulsion systems. Another is thermal insulation on the louvers of communications satellites where intermittent exposure to sun and darkness requires the responsiveness of the foam. Still another characteristic is its minimal outgassing. Therefore, contamination of delicate equipment and space sensors is one less worry. Solimide polyimide foam has been qualified for placement on Marshall Space Flight Center's materials specification selection list.

The foam technology is covered by several patents, some of them owned by Inspec Foams and others waived by NASA for Inspec Foams use.

Inspec Foams has established a manufacturing process that permits the material to be uniformly foamed. Solimide foam can be fashioned to carry an adhesive backing, exposed by peeling off an outer skin. A variety of densities and structural configurations can be produced, with the foam remaining resilient under exposure to temperatures ranging from minus 300 to plus 500 degrees Fahrenheit. Virtually no smoke or toxic byproduct is generated upon decomposition. Furthermore, the product's flame retardant qualities have prompted their application in ships and surface transportation systems, such as cars, trains, buses and automobiles.

The foam's properties also offer tremendous benefit in conventional insulation settings. It has good thermal and acoustical insulation properties; and the product is easy to fabricate and install.

There has been an outgrowth of uses for Solimide foam, stemming from the NASA work. Similar in need as that of the space agency, the U.S. Navy sponsored research by Inspec Foams to make new insulation systems for surface ships and submarines. Industrial uses include insulating supercold

fluids on tanks, pipelines, and wind tunnels that apply the foam as expansion joints.

Inspec Foams has developed Solimide foams to give airframe manufacturers major weight savings, while retaining their good thermal and acoustical properties. For example, when used in place of fiberglass in the lower lobe of a 747, the foam can generate a weight reduction of over 400 pounds. Solimide foams allow aircraft manufacturers to improve the designs of their insulation systems at an overall reduction in installed and life-cycle costs. That also means fuel cost efficiency. Aircraft applications include cockpit and cabin insulation, ducting, bulkhead insulation and under floor insulation. Using the product are such aerospace giants as Boeing, Lockheed, and General Dynamics, along with international clients like Airbus, Bombardiar, and Dornier.

A wide range of the high-performance Solimide polyimide foams is presently available.

®Solimide is a registered trademark of Inspec Foams Inc..

Initially developed for the Space Shuttle, polyimide foams are now in use in industrial and commercial applications worldwide.

Space age polyimide foams can be shaped for easy installation and laminated with barrier film. The foam offers uncommon properties as an insulation material.

Polyimide foam created for the Space Shuttle program can be fabricated for an assortment of industrial and commercial uses.

MOTION OF A MOVING OBJECT

The knees of a crash test dummy and the controls of a high-performance test aircraft. One might be hard pressed to think of a connection, but there is one.

Technology built for space-based and experimental aircraft programs is now in use in industrial settings and automotive testing, as well as auto and motorcycle racing.

SpaceAge Control, Inc. of Palmdale, California was awarded a contract in 1970 to support flight test programs at NASA's Dryden Flight Research Center, situated at Edwards Air Force Base in California. Their job was to produce precision, small format and lighter weight position transducers for aircraft flight control testing. The role of a position transducer is to convert mechanical motion into an electrical signal that may be metered, recorded, or transmitted. They are used in a broad range of position, displacement, and velocity measurements. Output from a position transducer can be measured to discern the position, direction, or rate of motion of the moving object.

The first application of the company's miniature position transducer was in support of a NASA project to study the actions of an aircraft's rear stabilizer. Following this work, SpaceAge Control was awarded a NASA contract to fabricate devices for obtaining angle of attack, sideslip, and pressure information on test aircraft. Since that time, the company has designed similar devices in the form of air data booms, nose booms, or air data probes, for an assortment of flight test activities.

Demands by NASA and a number of U.S., Canadian, European, and Asian aerospace firms have yielded a complete line of SpaceAge Control state of-the-art, miniature and subminiature-sized position transducers.

Durability of the devices has proven to be yet another key attribute. When used properly, certain classes of the position transducers can outlast the vehicle being evaluated. Some models are rated to exceed 50 million cycles of operation.

One Indy auto racing team began tasking the firm's position transducers to monitor the engine throttle movement and suspension travel of their racers. Similarly, NASCAR race vehicles have used position transducers to measure the rear spoiler deflection during high-speed test runs. By quantifying the amount of spoiler deflection occurring, race engineers can stiffen the spoiler accordingly, thus reducing deflection to an acceptable level. Any spoiler deflection results in less downforce and a higher riding car with a slower top speed. A variety of position transducers can be tapped to unobtrusively record driver throttle, brake, clutch, and gear shift actions.

SpaceAge Control products have been useful in a number of auto test and measurement projects. One such assignment has been measuring the brutal forces on occupants as they are thrown about during vehicle crashes. Built into the knees of anthropomorphic dummies, special transducers measure dummy knee shear. This effect occurs when the tibia (shin bone) moves relative to the femur (upper leg bone). Important leg-related data have been accumulated by incorporating transducers into the crash test dummies. The devices convert mechanical leg movement into electrical signals for detailed analysis. Position transducers have also proven helpful in monitoring displacements of a dummy's thoracic cavity and rib cage area as they undergo the jarring motions of an auto accident.

From their early work supporting NASA research, SpaceAge Control products have been successfully applied to military fighter, attack, and cargo aircraft, helicopters, and in general aviation craft. In the ground vehicle and transportation area, the company has customized devices for passenger cars, earth moving equipment, bus and public transport, as well as recreational vehicles. Now providing products to hundreds of customers in twenty industries and over thirty countries, the company can boast of one more measurable property: success.

Small, lightweight, and durable transducers provide a range of position and velocity measurements. The devices have proven helpful in the air and ground transportation sectors.

ROUTING THE POWER

The high demands of the space program for electrical relays have prompted the development of a quality product that now finds wide commercial application.

Kilovac, a division of CII Technologies based in Carpinteria, California, specializes in high performance, high voltage relays and contactors serving the aerospace, communications, medical, military and automotive industries.

Why are electro-mechanical relays so critical? Relays are electrically operated switches that can be located at remote locations. They are used to control electrical current or signals while providing electrical isolation. Electromechanical relays are designed to meet exacting circuit and environmental conditions while controlling numerous circuits simultaneously.

Innovations in switching technology were spurred, in part, by the challenging requirements of the International Space Station. Originally designed for the Lewis Research Center in Cleveland, Ohio, super-sealed switching technology enables a high degree of safety, independent of harsh environments, and permits a completely powered-off state during stops.

Kilovac has a proven track record aboard several space projects. The firm's devices have been utilized on the Telstar IV satellite, for controlling thruster firings and used in shuttle operations numerous times. In one shuttle application, Kilovac's relays are used in the main power supply for the shuttle-to-station docking system. SPAR Aerospace of Canada employs Kilovac contactors in their Special Purpose Dexterous Manipulator (robotic arm) designed for the International Space Station.

Another interesting space project that will haul Kilovac hardware beyond Earth orbit, to both a comet and an asteroid, is NASA's New Millennium craft, Deep Space One. This spacecraft is powered by ion engine, with Kilovac technology assuring the safe routing of high voltage from the probe's solar arrays to the ion engine.

Company skills in space projects have found other avenues of use, specifically in light rail electric train application, known more commonly as next generation people movers.

Kilovac is already a leader in electric vehicle switching technology, a market that is growing not only the automotive market but also the electronic Mars transit walker. For example, there are a number of cities that are considering electric mass transit systems to ease-up freeway congestion and lessen air pollution.

San Francisco's Bay Area Rapid Transit (BART) system is an early generation people mover. Pittsburgh and Atlanta airports also employ similar types of electric mass transit. The concept allows many fewer stops at stations since some of the cars can go directly to their occupant's destination without stopping at all possible stops in between.

Kilovac is working hard to design new and improved applications for these people movers in order to tap into this burgeoning market.

Relays fabricated by CII Technologies' Kilovac division switch main power inside each of the system's cars and control power routing during acceleration and braking. During braking, the energy can be put back onto the grid, thus reclaiming this energy instead of wasting it in brake pads and heat. Also, the large power interrupt of the Kilovac relay—as much as a 2,500 amp break—assures occupant safety even in cases of a component failure on board the people mover car.

Thanks to Kilovac's experience gained from association with NASA needs, the firm became involved in producing high reliability relays for the submarine cable industry. Just as in space, ocean application demands extremely high reliability and long life on the relays. Kilovac's special processes used in many relays now orbiting Earth were applied to the underwater relay design.

One of the processes implemented was small particle inspection. This process, also known generically as Millipore, is a way to ultra-clean a relay and void any particulate matter that could hinder the relay's performance over time.

Presently, a number of fiber-optic submarine communications networks are in progress or being proposed. These systems will link countries and continents to help shape a modern international communications network.

High reliability relays, having a performance rating of 25 years, are a must for any undersea telecommunications application. Kilovac is a lead supplier of the necessary relays to bring such a network to fruition. These relays must route high voltage along system segments, at very low leakage rates. Undersea cable elements can stretch from hundreds to thousands of miles long.

Kilovac relay switches have found a home in satellites as well as undersea fiber optic communications networks.

COMPUTERIZED ENGINEERING

The pervasive use of computers throughout the field of manufacturing has meant moving products more quickly to market, saving costs, and enhanced quality and performance. Computer-aided engineering has been a boon to the transportation industry, enabling the building of stronger and lighter aircraft, spacecraft, ships, and automobiles. This essential engineering tool has its roots in NASA research.

In the 1960s, Goddard Space Flight Center engineer, Thomas Butler, conceived a new dimension of mathematical precision in evaluating stress in complex structures. Under contract to NASA, the MacNeal-Schwendler Corporation (MSC) of Los Angeles, California developed an operational software version of, what was dubbed, the NASA Structural Analysis (NASTRAN®) program. Early adoption of NASTRAN proved of great value to NASA centers across the nation. NASTRAN later became available through the Computer Software Managment and Information Center (COSMIC). In 1973, the emergence of NASTRAN into the automotive industry was fortuitous, as an oil embargo triggered a flurry of research into lightweight cars, rapid transit vehicles, and fuel-efficient engines.

MSC's co-founder, Dr. Richard MacNeal, and his colleagues procured the rights to market versions of NASTRAN in 1982, gearing the software to an expanded industrial role. Subsequently, MSC proprietary versions of the computer problem-solver tool have become a mainstay for a diversity of applications, from heat transfer to acoustics.

In January 1995, MSC signed a cooperative agreement with the Langley Research Center for the inclusion of NASA's Finite Element Interface Technology into MSC/NASTRAN. Under the three-year agreement, MSC has received Langley-developed technology for modeling structural systems. This transfer of technology to non-aerospace companies from Langley is viewed as important to developing U.S. competitiveness in a global market. In July 1997, MSC announced the latest version of MSC/NASTRAN for Windows.

Over the years, MSC has grown to become the world's largest provider of mechanical computer-aided engineering (MCAE) strategies, software, and services. The company is at the forefront of providing finite element analysis (FEA) products.

FEA is best described as an analytical method used by engineers to help determine how well structural designs survive in actual conditions, such as loads, stress, vibration, heat, electromagnetic fields, and reactions from other forces.

Using FEA, virtually any structure, no matter how complex, can be divided into small elements. Those smaller elements form a finite element model that simulates the structure's physical properties. The model can then be subjected to rigorous mathematical examination, with an engineer able to see results in any number of formats.

Ultimately, FEA significantly reduces the time and costs associated with prototyping and physical testing. That testing can involve a diversity of products, from coffee makers to golf clubs and electric guitars to the Space Shuttle. MSC/NASTRAN, the firm's flagship product, can model and analyze almost any material, including composites and hyperelastics.

Another major product from MSC is MSC/PATRAN, which was also derived from early NASA work. MSC/PATRAN provides an open, integrated, MCAE environment for multi-disciplinary design analysis. This powerful tool integrates computer-aided design, test, and analysis software. It can be used early in the design-to-manufacture process to simulate product performance and manufacturing processes, giving the user a way to visualize and interpret data with new understanding. In October 1997, the Japanese firm, Denso, the world's largest auto parts supplier, adopted MSC/PATRAN as a core product in that company's 3-D and virtual prototyping work.

For three decades, the unique prowess of MSC/NASTRAN has proven itself time and time again. It has permeated the development of structures and systems, advancing the art of transportation. This computer application, having thousands of users worldwide, has played a key role in maintaining swift and steady progress in a dynamic technological arena.

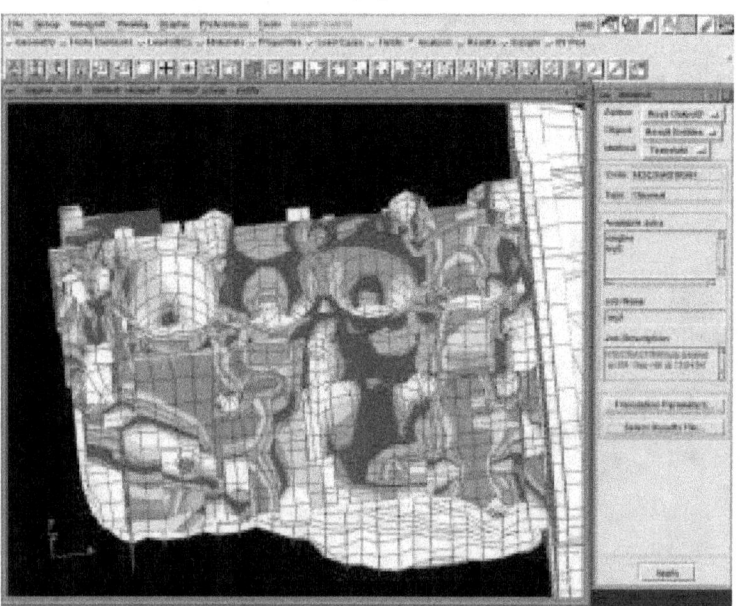

Stresses and strains of this engine are displayed by using MSC/NASTRAN computer software.

®NASTRAN is a registered trademark of NASA.

SOFTWARE SMARTS

Planning a space mission is a complicated matrix of people, hardware, and flight objectives. Knowledge Based Systems Inc. (KBSI) of College Station, Texas, designed project management software that allows NASA project managers to script activities for the Space Shuttle. The work was funded through a Johnson Space Center Small Business Innovation Research (SBIR) contract. The Mid-Continent Regional Technology Transfer Center also helped KBSI locate federal funding to later commercialize the software, permitting it to move from demonstration level into production.

In commercializing the NASA-funded product, KBSI was then afforded the opportunity to further modify the software, making it an add-on to a popular process modeling and simulation software.

The software's analytical tools give users a way to gauge the impact of scheduling, monitoring, and managing subcontractors and suppliers. Cost variances in projects can be predicted and detected. Additionally, the accuracy and consistency of cost estimates are improved.

Certainly an invaluable virtue of the KBSI software is understanding relationships between people, tasks, and costs related to a project. By capturing lessons learned, the software grows in value over time.

One early software success story involves a multibillion-dollar service provider. The company needed its new-hire process redesigned. A previous procedure took seven weeks to place a new employee on the job. That resulted in long cycle times, poor-quality products, and unhappy customers.

Using KBSI's planning software, the key issues brought about by redesigning the company's new-hire process could be mapped out, step by step. How best to execute the redesigned procedures and the appropriate cost models to enact those steps were established. Results were immediate. The new-hire process virtually eliminated processing errors, reducing system requirements by ninety percent. Once implemented, the company started to enjoy an entirely redesigned new-hire process—one that has a fresh employee processed and placed in just one day.

In another instance, for a consortium of major U.S. semiconductor manufacturers, the software developer provided computer tools to model, design, and optimize manufacturing processes. An advanced system to capture expert knowledge for design-to-cost analyses was also developed.

KBSI-developed software is also in use by a major American automotive company. With KBSI expert systems in place, the auto maker applied the tools to design car air conditioning and cooling systems.

Founded in 1988, KBSI's initial mission was forward thinking: to advance the state of the art by developing creative solutions for a wide range of critical problems faced by government, academic, and industrial communities. The company has since evolved to set commercial and defense industry standards for the development and support of modeling, and analysis tools and methods.

KBSI's software products may be used as stand-alone tools, or to automatically share information with a growing list of other software vendors. Company customers include Lucent Technologies, Comprehensive Technologies International, the Chrysler Corporation, and a slate of government agencies.

The credo for Knowledge Based Systems Inc. is direct. Take control of how you do and what you do. Prepare contingency plans for when the real world waylays your best-laid plans. Manage risk by doing business better, faster, and smarter. In the end, KBSI's software promises its user that you can plan today for tomorrow's success.

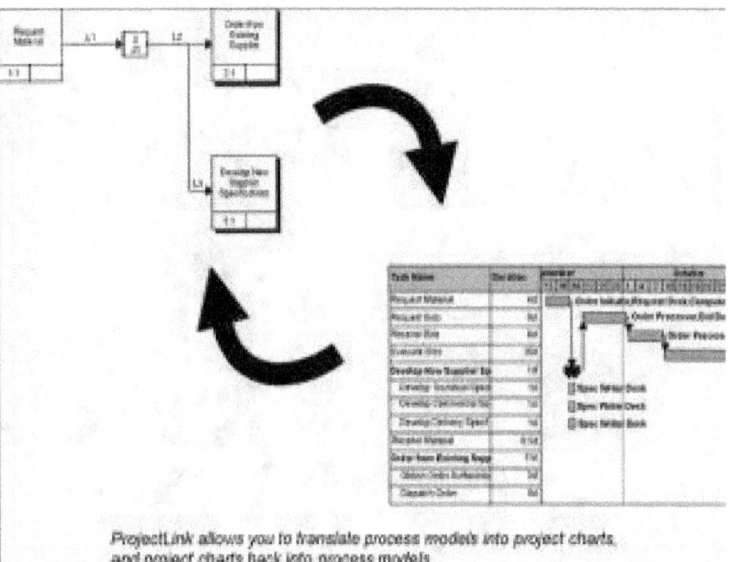

ProjectLink allows you to translate process models into project charts, and project charts back into process models.

KBSI's project management software, allows you to translate process models into project charts, and project charts back into process models.

IMPROVING AIRLINE SAFETY

Three-dimensional (3-D) simulation software for virtual reality applications has found new utility in the airline industry. Thanks to the efforts of Steve Lakowske, founder of SimAuthor Inc. of Boulder, Colorado, a flight data visualization system called FlightViz™ has been created for NASA's Aviation Performance Measuring System (APMS), resulting in a comprehensive flight visualization and analysis system. The visualization software is now capable of very high-fidelity reproduction of the complete dynamic flight environment, including airport/airspace, aircraft, and cockpit instrumentation.

The APMS program calls for analytic methods, algorithms, statistical techniques, and software for extracting useful information from digitally-recorded flight data. APMS is oriented toward the evaluation of performance in aviation systems, particularly human performance, Lakowske explains.

Lakowske is a ten-year NASA veteran, formerly employed at the Ames Research Center. He attributes much of his success to the years of NASA work in computational human performance modeling. His NASA experience in simulation technology began at Ames in 1979, working as an electronics engineer in the Electro-Systems Engineering Branch.

SimAuthor software features textured 3-D scenes and high-fidelity reproduction of an aircraft's complete dynamic flight environment.

Flight data are routinely analyzed by many non-U.S. carriers to measure the safety, training, and efficiency of their fleet operations. There are many ways of using digital flight data for these purposes. One of the specific goals of the APMS research project cites the need for enhancing and facilitating animated playback of data from individual flights to assist data analysts, and to provide effective feedback for training and self-assessment of air crews.

The innovation prompted by the APMS work centered on the fact that current, commercial-off-the-shelf technology to animate flight data was designed for accident investigations, not for crew feedback. The fidelity of this animation, therefore, was not satisfactory for the objectives of APMS.

Formed in 1996, Lakowske's SimAuthor Inc. is a Colorado corporation established to engineer turnkey systems for authoring simulation-based analysis and training systems. "The objective of our company," adds Lakowske, "is to enable a broad range of non-programmer professionals to quickly and easily create sophisticated, dynamic 3-D, high-fidelity, interactive simulation systems without requiring computer programming expertise.

Lakowske is quick to point out that multimedia tools and techniques are often used today to address a wide variety of training and corporate communications needs. But even these high-tech software tools can be enhanced, he says. Two additional enabling technologies can extend the virtues of multimedia approaches: High-fidelity, fully-articulated, interactive, 3-D real-time computer-generated graphics; and high fidelity, real-time physical device simulation modeling.

In fulfilling certain goals of the APMS effort and related Space Act Agreements, SimAuthor delivered to United Airlines in 1997, a state-of-the-art, high-fidelity, reconfigurable flight data replay system. The software is specifically designed to improve airline safety as part of Flight Operations Quality Assurance (FOQA) initiatives underway at United Airlines, as well as Advanced Qualification Program (AQP) or training requirements.

Pilots, instructors, human factors researchers, incident investigators, maintenance personnel, flight operations quality assurance staff, and others can utilize the software product to replay flight data from a flight data recorder or other data sources, such as a training simulator. The software can be customized to precisely represent an aircraft of interest. Even weather, time of day, and special effects can be simulated.

Among the key benefits of the software product is its ease of use. Drawing from simulator databases of airports and public domain cockpit instrumentation, the software lessens user training time due to intuitive graphical user interfaces.

Now adopted by United Airlines, SimAuthor's state-of-the-art flight data replay system is enhancing safe, day-to-day operations of aircraft and bolstering confidence in air travel by the flying public.

™FlightViz is a trademark of SimAuthor Inc..

ELECTRIFYING DEVELOPMENT

It can happen in a split second. A lightning bolt strikes your home, sending a destructive pulse of energy through electrical wires. Expensive electronic equipment, such as satellite dish systems, antennas, television cable hardware, and even sprinkler systems can be damaged. The statistics are shocking in themselves. A hundred times per second of every minute of every day, lightning strikes the Earth. One outcome is over two billion dollars worth of damage, caused each year by lightning in the United States alone.

A defense against lightning has been invented, brought about with assistance from Kennedy Space Center and Marshall Space Flight Center engineers.

This success story began to emerge in 1979, with inventor Sam Gasque of Flat Rock, North Carolina, creating the all-in-one cable for home satellite systems. His idea was to take the many loose wires used to power the satellite dish and package them into one single, neat, direct-buriable cable.

However, reports of damaged satellite dish equipment due to lightning strikes prompted Gasque to redesign his product. That redesign led to lightning retardant cable, a

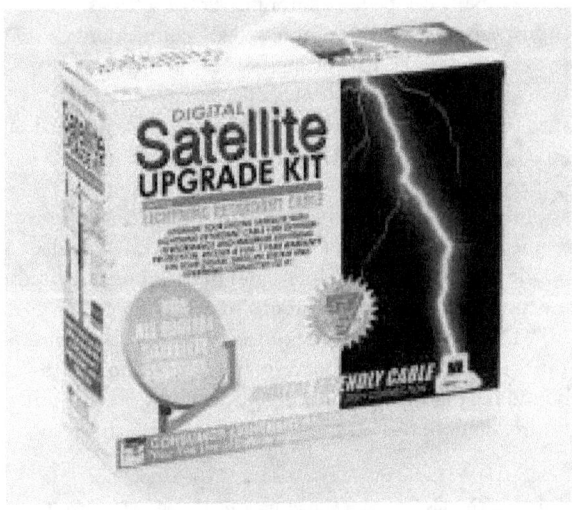

Consumer upgrade kits for protecting small satellite dishes benefited by research on preventing lightning strikes to the Space Shuttle and its launch pad.

product that works in part to cancel magnetic field effects generated during a lightning strike. Gasque contacted engineers at the Kennedy Space Center, seeking help to validate his revamped invention. Further fine-tuning of the design made it all the more functional.

In a quest for further help, Gasque turned to the NASA-Southern Technology Application Center (STAC) at the University of Florida. Through the NASA-STAC, Gasque was able to review research on lightning protection of the Space Shuttle and its launch pad. Gasque used STAC resources to also look for information on all worldwide patents on lightning protection involving

cable. Over a foot-and-a-half high stack of information about the topic was compiled.

One search result led to Lightning Technologies, Inc., of Pittsfield, Massachusetts, a group who had previously done lightning tests on the Space Shuttle. Gasque contracted the firm to test his lightning retardant cable, an evaluation which showed his product offered between a 700 percent and 1,480 percent improvement over standard cable.

Gasque's tie to NASA and Space Shuttle research helped validate and improve the lightning retardant cable, making it all the more marketable through his company, Consumer Lightning Products/GS Cable, Inc., situated in Flat Rock.

The lightning retardant cable has since been installed in several homes on a South Carolina island that has a history of frequent lightning strikes. The cable's performance was monitored for several years, during which it took several direct hits. Gasque said not once was lightning carried into the homes through his cable innovation.

After an exhaustive campaign of research and development, the cable product has entered the consumer market. Inroads are being made in the commercial industry, specifically in airport lightning systems. Successful installation of the lightning retardant cable can also provide surge protection in power lines, telephone lines, and off-air antennas or cables. As for satellite dish systems, in all cases where lightning struck an installation, the properly grounded lightning retardant cable prevented the equipment from being damaged—even in situations where telephones and televisions were destroyed.

Pre-packaged kits of the protective cabling are being sold. Consumer Lightning Products has employed the handicapped and physically disabled to assemble and bundle up the do-it-yourself kits. In the hands of a customer, installing the cable system to a homeowner's satellite dish and receiving equipment is simple, taking approximately one hour.

First year sales, reports Gasque, are expected to exceed $10 million, with a volume of $200 million by the fifth year. The manufacturer hopes to establish a new manufacturing facility in northern Georgia or western North Carolina to accommodate demand.

Lightning protection cable system can thwart the destructive pulse of a bolt out of the blue.

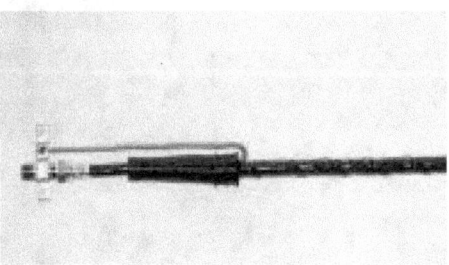

SECURE FUTURE

In the business of space construction, the phrase, "some assembly required" is commonly heard, just as it is here on Earth. But there is a big difference. Suited up in heavy pressurized space suits and outfitted with bulky gloves, astronauts have a tougher job in completing work tasks.

Specifically built to Johnson Space Center requirements, Thread Technology, Inc. of Sterling, Virginia, developed the ZipNut®, replete with Push-on Threads®. As the name suggests, this fastener can be pushed on, rather than turned. Originally developed for the Space Shuttle and space station programs, the product is now in use by firefighters, nuclear powerplant repair technicians, and others involved in difficult assembly duties. These quick-connect fasteners enjoy the flexibility and strength of threads while eliminating their heretofore inherent weaknesses of slowness and cross-threading.

NASA has adopted the ZipNut fastener for space walking and robotic in-space assembly. A tool to apply the fastener was first developed for flight aboard a Space Shuttle in 1989. It was also adopted for the Space Station in 1992. The attachment technology has been a part of two Hubble Space Telescope repair and servicing missions, in 1994 and 1997. Zipping on and zipping off attachment handrails to move delicate Hubble instruments about the Shuttle's cargo bay were made possible by using the exceptional fasteners.

Once segments of the International Space Station are orbited, astronaut "hard hats" will be faced with piecing together the various elements. Thread Technology is supplying ZipNuts to help assure fast and easy connection of Space Station hardware. Installation time can be improved since bolts can be pushed into place rather than having to be turned like conventional nut/bolt combinations.

There are several benefits and features of Thread Technology fasteners, also making them ideal for more down-to-Earth applications. Coupling to an existing threaded part can be done with only a push and twist of up to one full turn. Wrench tightening is obsolete. Rapid assembly is possible without the need of starting the thread lead or rotating the male and female parts multiple times.

A patented optional feature of the ZipNut allows quick removal of fasteners and connections. A locking system prevents accidental disconnection under load. For many markets, such as industrial gas filling, this feature reduces repetitive hand motion problems and virtually doubles the product benefit without doubling the cost.

The Push-on Thread fastener also resists movement under vibration. Adverse environmental conditions do not degrade product performance, whether exposed to outer space, contamination, or extreme weather. Lastly, the product can be fabricated from most materials, such as stainless steel, aluminum, titanium, brass, and even plastic. There is no material limitation so long as the material has sufficient strength to bear the specified loads. A fastener can be designed in a wide range of sizes, such as the one-quarter-inch nut developed for NASA or a six-inch high pressure water coupling.

ZipNuts can be found in numbers of settings: Industrial and medical compressed gas connections; hydrant and water hose fittings; for extracting a person from a crashed auto, building collapse, cave-in or construction accident; and in radioactive areas where speed of attachment and removal of threaded items are a must. Other potential product markets include automotive assembly and operation such as lugnuts, coaxial and pin connections for electronic equipment, injection-molded plastic parts for mass assembly, and in high pressure fluids for fire or dealing with hazardous materials.

Thread Technology envisions ever wider use of the ZipNut in future years, not only in space but for Earth-bound, in-the-home chores, too. It will be a "fasten"-ating future, no doubt!

®ZipNut is a registerd trademark of Thread Technology, Inc..
®Push-on Threads is a registerd trademark of Thread Technology, Inc..

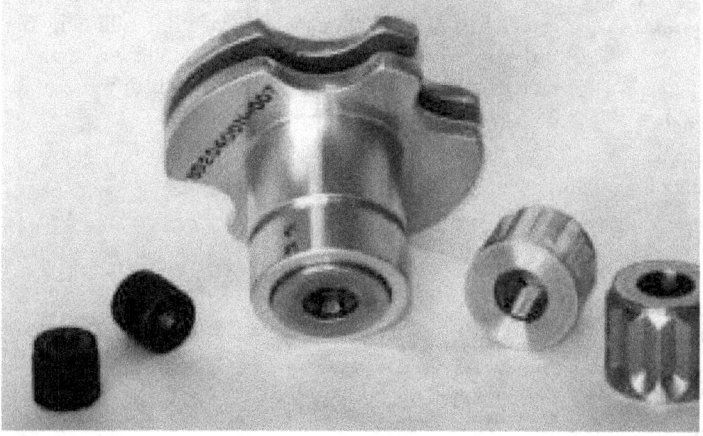

ZipNut® fasteners are unique in that they push on rather than turn. NASA has adopted this technology for space walking and robotic in-space assembly.

EYEING THE INVISIBLE

Software tools to monitor vapor concentrations using Fourier Transform Infrared (FTIR) spectroscopy received an assist by way of a nonreimbursable Space Act Agreement between the Kennedy Space Center (KSC) and the MIDAC Corporation of Irvine, California.

The agreement which was initiated in 1995, was established to encourage software improvements to an FTIR spectrometer. This vapor monitoring hardware is on duty at KSC's Space Station Processing Facility. Ammonia is called for in the current International Space Station design, to be used as a refrigerant. Ammonia must be closely monitored during loading and storage due to its toxic and flammable nature.

Infrared technology, like the FTIR, allows the invisible to be seen. Invisible vapors can be measured by this technology because each compound absorbs certain frequencies in the infrared spectrum. These absorbed frequencies uniquely identify each compound, like a fingerprint.

Prior to the collaboration between KSC and MIDAC, one of the company's instruments was in demand, utilized in contamination and toxic vapor detection studies. Sample measurements would be churned out in raw form by the equipment. This basic data would be fed into an external computer. It would transform the raw interferogram information into information concerning what compounds, and how much of each, were being measured in the sample.

Kennedy Space Center lab technicians installed a small computer inside the MIDAC FTIR instrument. The KSC specialists also developed the software code for calibrating the monitor to recognize and measure a given infrared-active gas or vapor. A runtime code for operating the vapor monitor was devised. This augmentation of the FTIR, outfitted with an embedded computer and special software, could then produce a list of compounds and their concentrations, instead of the raw unprocessed data.

A number of troublesome factors were eliminated by the new software coding, such as temperature drift of the FTIR, the effects of optical degradation. Improved sensitivity of the device was attained, despite wide fluctuations in environmental conditions during system operation. Data stream output to a communications port was produced in an easily readable format.

The Space Act Agreement allowed for these codes to be polished into a commercial grade product, of value to both NASA and MIDAC. That commercial software package has been furnished to NASA and is also now incorporated in the MIDAC line of products.

KSC's Contamination Monitoring Laboratory designed, fabricated, and delivered a Portable Ammonia Monitoring System with the new software. Called the Ammonia Detection Cart, it has been used during validation testing of the Ammonia Servicer in the Space Station Processing Facility. This servicer contains and controls the ammonia loaded into space station elements. The final versions of the cart have been delivered for environmental area monitoring during the processing of Space Station elements that are to carry liquid ammonia. Capabilities of the cart have also been demonstrated for alcohols, Freons, ketones, water vapor, carbon dioxide, and the like.

MIDAC's FTIR technology has proliferated in use over the years, particularly in industrial process applications where emission rates must be measured. MIDAC offers a range of rugged, high performance spectrometers and components to customers. An arsenal of analytical tools has been crafted for virtually any gas analysis needed, capable of working in real-world environments. MIDAC instruments are built to provide fast and easy access to accurate data.

MIDAC systems are providing critical data in a variety of areas including: environmental monitoring, process optimization and control; applications for semiconductor manufacturers; fenceline monitoring of industrial facilities; rocket plume analysis; and studies of volcanic emissions.

"Fingerprinting" types of chemicals and their amounts permits effective emission monitoring, protects the environment, ensures health and safety of workers, and provides cost savings in the manufacturing process.

IDENTIFICATION SYSTEM

The parts count for a Space Shuttle is a whopping number in the millions. Take for instance a Shuttle orbiter's protective veneer of heat shield tiles. Each of the thousands of tiles is catalogued as to size, type of tile, and location on the vehicle. To identify, track, and record keep these tiles, digital data matrix technologies were developed at the Marshall Space Flight Center.

Today, individual heat shield tiles on the fleet of orbiters are marked with a high data density, two-dimensional, machine-readable symbol. This NASA-developed technology helped to launch a new commercial endeavor in product coding.

Through a NASA Space Act agreement, CiMatrix Corp. of Canton, Massachusetts has been able to commercialize the product coding idea by establishing the Symbology Research Center (SRC) of Huntsville, Alabama. Opening its doors for business in August 1997, SRC is commercially marketing a new method of identifying products with invisible and virtually indestructible markings. These laser-etched markings are termed "compressed symbologies."

SRC offers compressed symbologies as a way to automate inventory and cut warehousing costs and avoid part shortages. Other benefits of direct parts marking are updating the part's history in real-time, increasing read rates to virtually 100 percent, guaranteeing part/component integrity, and eliminating paper labels and tracking paperwork.

No longer does a company have to face missing paper labels—labels that can fall off a high-value part or product due to heat, cold, rain, wind, and other inhospitable conditions.

The permanent digital data matrix codes work on practically any surface, be it steel or metal, even plastics, glass, paper, fabric, ceramics, or other material. Compressed symbologies can withstand extreme fluctuations of temperatures, up to 2,200 degrees Fahrenheit and an air flow exceeding 18,000 miles per hour. That is the heat load and speed associated with a Shuttle orbiter during a space mission.

The coding technology provides up to 100 times as much information as linear bar coding symbology in the same or less space. Markings can range

Permanent direct marking of products assures ease of tracking and record keeping.

in size from a mere four microns (read microscopically) to as large as two square feet (read telescopically). Commercial uses for the new labeling technology abound. According to Donald Roxby, SRC's Director, the marking system is attracting a varied clientele with a range of commercial interests. "Everything from electronic parts to pharmaceuticals to livestock," Roxby reports. He sees a growing need for an identification system that can be placed directly on a product, regardless of that product's shape, size, color, and other features.

Additional commercial markets for the marking technology have been targeted. These include counterfeit fasteners that can be discovered and removed without replacing every fastener on the product. Engine parts can be accurately tracked during the product's entire lifetime. Delicate and expensive items can be easily and more accurately inventoried and guarded against theft.

The growing list of customers for laser-etched symbol coding already includes major companies in the automotive business, such as General Motors, TRW, and Borg Warner. In the electrical and semiconductor business sector, Hewlett Packard, Intel, and Motorola are customers. Pharmaceutical suppliers Johnson & Johnson, Eli Lilly, and Parke Davis also use symbol coding on their products.

Quick solutions for overcoming marking problems are available from the Symbology Research Center. Considered the most advanced 2-D symbology research and development laboratory in the world, SRC maintains the country's most comprehensive materials marking database. The center has an ongoing Space Act Agreement with NASA to further advance this 2-D technology. Any governmental or commercial entity can request assistance on a specific product identification problem by submitting a problem statement through the Marshall Space Flight Center Technology Utilization Office or directly through the SRC.

Director of Symbology Research Center, Donald Roxby, inspects laser-applied product labeling.

HIGH-TECH, LOW-TEMP INSULATION

The fiery reentry of future reusable space planes can receive a cool reception as they slam into the Earth's atmosphere by using a new lightweight metal insulation. That same technology has been applied to the creation of emergency rescue blankets and mittens capable of thwarting extremely cold weather.

S.D. Miller & Associates of Flagstaff, Arizona has received Small Business Innovation Research (SBIR) contracts through the Ames Research Center. The company is operated by Steve Miller, who has served as the principal investigator on several NASA SBIR awards over the past 11 years.

The SBIR work has launched an investigation into a unique flexible insulation blanket suitable for the thermal protection systems of future spacecraft during atmospheric entry. A low-density, honeycomb-like material was fabricated, capable of inhibiting convective and radiative heat transfer. This advanced, but lightweight, insulation was made from special metal alloys and ceramics. Shaving off any weight from a reusable launch vehicle means a decrease in fuel and frame weight and, ultimately, lowers the cost of hurling each pound of payload into space.

Spacecraft are not only vehicles to benefit from using the improved lightweight, multi-layer thermal insulation. Current aircraft designs can be made more efficient by reducing the weight of the insulation system, such as that used in certain areas of a jet engine. The cloth-like honeycomb material can withstand temperatures as high as 2,200 degrees Fahrenheit. An outstanding characteristic of the fabrication process is that a broad range of materials can be used to form the honeycomb to match the temperature range of the application.

Miller's group saw several potential spinoff applications using the same honeycomb concept. Collaborating with NASA scientists, the team made a lightweight plastic insulation for blankets and clothing that has properties better than wool fibers or polyester fleece. Better yet, the honeycomb structure can even be made from recycled milk containers—an excellent use of refuse.

Utilized in blankets, the plastic honeycomb material keeps a person four times warmer than wool, even when subjected to cold and wet conditions. Blankets made from the plastic insulation are also non-allergenic, and dry five times faster than blankets made of wool. Thermalon Industries, Ltd. in El Segundo, California, is commercializing the technology in various products including blankets and mittens. "Our first production run of mittens was sold out before New Year's. We believe they were bought mostly by extreme skiers and ice climbers through word of mouth advertising," says Miller, who is a principal stockholder in Thermalon.

Eventually, about 70,000 emergency blankets that use the new insulation technology are expected to be distributed annually by Thermalon Industries. A program to demonstrate the unique attributes of the blanket was begun with NASA Ames' Disaster Assistance and Rescue Team. Also planned is a full evaluation of the blankets in concert with ambulance companies and American Red Cross chapters.

Yet another commercial product is emerging from the space program work. Mittens are now in production that are warmer than wool and made from recycled plastic.

The mittens are designed for people in extreme cold weather, including recreational, industrial, and military users. Informal field testing of the mittens has been carried out in the frozen climes of Antarctica, Miller reports.

Silk is used as a glove lining for greater breathability and extra comfort. A waterproof/breathable shell on the outside of the product facilitates faster drying. Glove palms use a rubberized material for better grip.

Since hands perspire more than other parts of the body, handwear often gets damp and this moisture greatly increases heat loss from the hands. "Our honeycomb insulation doesn't trap moisture, so hands stay warmer," Miller says. "Active people from snowboarders to driveway shovelers will benefit since perspiration increases with greater activity," he adds.

Other spinoff insulations are also deemed feasible. The Whirlpool Corporation is evaluating the material as a moisture-tolerant alternative to chlorofluorocarbon (CFC)-blown foam that would make their refrigerators even more efficient. Units that hold and maintain super-cold fluids, and other industrial applications would also benefit from the technology's unique advantages.

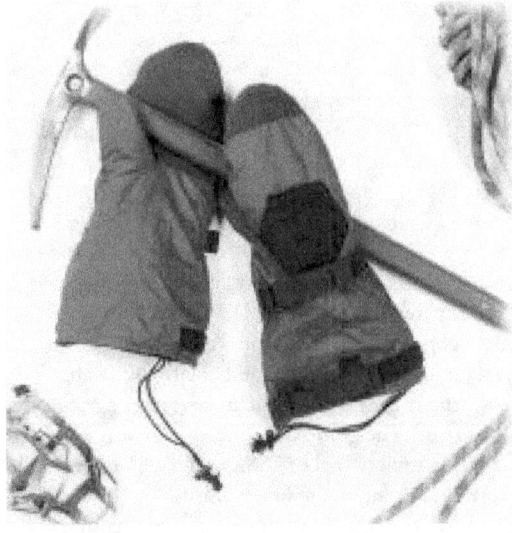

Material to protect reentering space planes is an ideal glove insulator for keeping hands warm in cold, harsh climates.

SLOW-RELEASE FERTILIZER

Zeolite already sounds unearthly. But as part of a gardening fertilizer, this group of minerals with an exotic name has earned the space age equivalent of a green thumb of approval.

NASA has long been interested in ways to sustain plant growth in space environments. Plants are being eyed as critical to prolonged space exploration, supporting astronauts with oxygen, food, water, and to help recycle waste products as part of a regenerative life support system. One problem, however, was the complexity of system hardware to maintain water and nutrients for plant growth. Hydroponic systems involve a set of circulating pumps, sophisticated sensors, and complicated control and monitoring systems. The search was on for a plant growth system that minimized such a hardware maze and eliminated "free" water which tends to float about in space. Additionally, the plant growth system had to operate reliably in space. Using a medium analogous to soil was a simplifying solution.

A highly productive, synthetic soil (or substrate) was created to support plant growth. This research led to the cultivation of plants in zeolite mineral substrates that contain essential, plant growth nutrients. The technique was termed "zeoponics."

Zeolites are a group of naturally occurring minerals having a honeycomb-like layered crystal structure. Its network of interconnected tunnels and cages can be loaded with nitrogen and potassium, combined with other slowly dissolving ingredients containing phosphorous, calcium, and a complete suite of minor and trace nutrients. The zeolite acts as a reservoir for the nutrients which are slowly released "on demand." Indeed, the plant itself does the regulating of the nutrients as it needs them. With only the addition of water, plants will grow in the zeoponic medium for multiple growth cycles.

Working with NASA's Johnson Space Center and the Kennedy Space Center under Small Business Innovation Research (SBIR) contracts, Boulder Innovative Technologies (BIT) began the initial work on the superior plant growth media for long-term space travel. ZeoponiX, Inc. of Louisville, Colorado was established as a sister and spinoff commercializing company from BIT to introduce zeoponic products. The SBIR-developed technology is being licensed by BIT to ZeoponiX, who also holds exclusive rights to issued patents from NASA.

Explains ZeoponiX's CEO and chairman, Richard Andrews, "the company has brought space age agriculture down to Earth." The group introduced its first product in 1997, a direct spinoff of the NASA-funded SBIR research, he adds. Other products are in the works.

The first creation is currently sold to golf courses, sports playing fields, and for greenhouse horticultural use. As a proven combination fertilizer and soil amendment, the mixture is showing its prowess in both construction and renovations of golf greens and tees, and toughening up the turf of playing fields. The product can be blended into other commercial and consumer potting mixes. Its use provides a lasting reservoir of nutrients allowing the user to reduce added fertilization while achieving better plant and vegetable performance. There is an environmental bonus from the product as well. Unlike commonly used fertilizers, the plant-growth material dramatically reduces loss of nutrients to groundwater and the environment.

A number of key technical advances and benefits of the company's zeoponic products include:

- a combination of a superior growth medium and soil conditioner, as well as a fertilizer delivery system;
- an all-in-one balance of major, minor and trace nutrients;
- high-efficiency delivery of nutrients to the plants;
- much lower losses of applied nutrients to the environment; and
- exceptional root development and improved plant performance.

"Several products are planned for release during 1998 and beyond, based upon this NASA-licensed technology," Andrews says. One will be a product for use by the homeowner/consumer, a potting mix additive, and possible pre-blended potting mixes containing the NASA-licensed ingredients, he adds.

ZeoponiX has an agreement in place with a firm in Australia to manufacture these products based on slow-release fertilizer and active synthetic soil in Australia and New Zealand. Active discussions are also underway with other companies in other countries, including Canada, South Korea and in Europe.

TM ZeoPro is a trademark of ZeoponiX, Inc..

The Ohio Clearview Golf Club used ZeoPro™ nutrients to change the type of grass on the green. Success has exceeded expectations—due to the release of zeoponic materials in the root zone of the new seedlings.

REAL WORLD AUDIO

Acoustic bliss—an interactive, real-life audio experience by surrounding the listener with sounds in three dimensions using only a single pair of ordinary speakers or headphones. Getting an *earful* earns an entirely new and enjoyable meaning thanks to the audio know-how of Aureal Semiconductor, Incorporated of Fremont, California, and its subsidiary, Crystal River Engineering (CRE), Incorporated of Palo Alto, California.

"We hear the future" is a behind-the-scenes slogan that drives the work. That spirit to advance audio technology was initially spurred by CRE's work in collaboration with the Spatial Auditory Displays Laboratory at the Ames Research Center. Very high-speed digital audio processing systems were developed to further Ames' research on virtual acoustic displays, part of the center's broader investigation into virtual reality and advanced multimedia displays for aviation and space applications. The NASA objective was to explore the possibilities of combining a three-dimensional (3-D) auditory system with visual virtual displays.

Aureal develops and markets a line of products that incorporate 3-D sound into video games, surround sound systems, computer sound cards, Internet sites, and other interactive software applications.

By furthering what is termed psychoacoustic research, audio standards have been raised to new dimensions. Aureal has embraced this research, creating innovative technology that alleviates the need for pre-encoding sounds or adding extra speakers to achieve "virtualized" experiences. The 3-D audio technology enables interactive placement of sounds in the entire 3-D space surrounding a listener. The advantage is obvious: A new generation of audio experience that is interactive, immersive, and fully three-dimensional.

Audio accelerators, for instance, can turn a computer into a thundering, true-to-life sound machine equal in quality to home theater surround sound systems. The result is a transformation of game playing into a visual and audio romp, immersing the user in a more interactive experience.

The interactive 3-D audio technology can be heard on the Internet. By using virtual reality modeling language browser software, Internet surfing now includes exploring 3-D websites. A user can be totally immersed in 3-D worlds, both visually and acoustically. As the user travels around a three-dimensional environment, sounds are rendered in their true locations with respect to the user. For example, if a virtual person is standing behind a user in a world, the person's voice will sound like it is projected from behind and will travel around the side and to the front as the user turns to face it.

Even a virtual sing-along, Karaoke style, has been introduced using the advanced audio effects technology. An effects processor chip offers voice pitch shifting, vocal accompaniment and vocal track elimination, among its functions. Fun vocal effects are also available, as is an application for people who are not comfortable singing in their own voice but prefer to amaze an audience with the warbling of a virtual singer.

Top software developers are enthusiastically embracing the 3-D audio technology to maintain an edge of competitors. Already signed up are such software companies as Acclaim, Activision, Electronic Arts, Epic, and LucasArts. The first titles using the audio advancement have reached store shelves, with dozens more in production. Technology licensing of the three-dimensional audio enhancement has permitted its use in an array of multimedia products offered by such developers as Analog Devices, ATI Technologies, Cirrus Logic, Diamond Multimedia, S3, Oak Technology, and Rockwell International.

Aureal has been able to bring 3-D audio from NASA's high-end research work in jet cockpit displays and flight simulators to mainstream electronic entertainment, consumer electronics, and communication applications. Applying that research has brought a level of awareness, realism, immersion, and engagement to the user, once only possible in real-life situations. By enveloping a listener in a three-dimensional sound field, a user is no longer aware of the audio system that is rendering the sounds.

Indeed, a new level of audio experience...and a sound investment in the future.

A high-speed digital audio processing system enables three-dimensional sound to be used in numerous consumer products, from computer games to home entertainment equipment.

WRAP-AROUND WORLDS

Cyberspace will never be the same. Vision, sound, vibration, and motion cuing come together to offer a sensory extravaganza—a personal encounter with virtual space.

Flogiston Corporation of Austin, Texas has designed a reclining relaxation chair, replete with a wrap-around, bubble dome display built into the head of the chair and a projector, as well as delivery systems for sight, sound, and motion. The result? A sensation of floating in space.

The product was fostered by a NASA Small Business Innovation Research (SBIR) contract managed by the Johnson Space Center. This NASA spinoff was first developed for astronauts as a virtual workstation, a space-like classroom to hone spacewalking skills. The chair incorporates research gained by NASA in designing spacecraft that carry human passengers, as well as data on the natural position a body assumes when floating in the microgravity of space. Using this posture information, Flogiston's patented reclining chair provides a minimal stress environment for a person interfacing with computer and electronic systems for extended periods.

The two-foot diameter dome offers the chair's occupant a complete field of view of virtual space. This projected cyberspace image wraps around the head of the participant, occupying the person's peripheral vision.

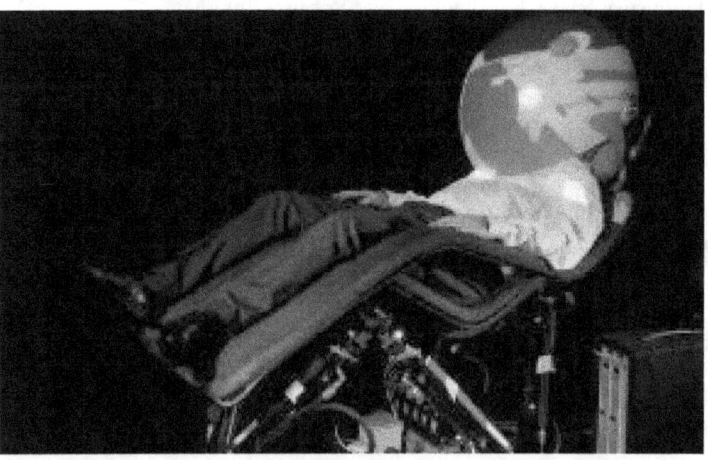

Instead of cumbersome head-mounted displays or gloves, the Flostation consists of a relaxation chair with a bubble dome display built into the head of the chair. The participant reclines in the neutral posture similar to astronauts floating in zero gravity.

Visuals are augmented by high-quality, surround sound speakers mounted inside the dome.

Shakers are built into the chair to provide whole body vibration, with higher-end chair models having enhanced vibration abilities. Mounted on a three- or six-axis motion platform, a generic motion simulator for simulated digital environments is created. Movements by the chair are synchronized with the operator's motion cuing actions inside the digital world. Left and right hand controllers are mounted on the arms of the chair. These hand controllers can serve several functions, such as navigating in space, or manipulation of objects in three-dimensional space.

Real-time video or recorded media can be accommodated, as can computer graphics, involving active or passive interaction. State-of-the-art audio and video hardware is utilized giving the chair's user a totally immersed—eyes, ears, body, and hands—virtual experience.

Flogiston president and design engineer, Brian Park, explains that a wide array of commercial applications for the full immersion chair are foreseen. Applications include entertainment, educational training, and stress management, even as the ultimate travel brochure for the prospective vacationer.

Motorola recently purchased one of the firm's basic models for testing and demonstrating new image warping technology they are developing, Park says. Intel, the computer chip manufacturer, has also purchased a chair for demonstrating immersion.

Demonstrations of chair models at medical, computer, and electronic trade shows have been well received, Park reports. Health community specialists saw the device as a tool for treatment of phobias, pre-operative education of patients, and as a workstation of the future for surgeons.

Numbers of chairs can also be operated in tandem, depending on computing resources available. Doing so can yield a shared immersion of a team that must collectively work together while performing individual tasks, or for networked gaming on the Internet. Clearly, the entertainment and education value of the chair for theme parks, be it real-time viewing of the Earth from space, undersea exploration, or flying over the terrain of a distant planet, is a targeted user community.

If immersive cyberspace is to be the natural method of computer interaction in the next century, Park says that his creation "is a significant step towards providing an effective and useful form of immersion for everyone." A low-cost, entry-level model of the chair without a motion base is being developed for home use, Park reports.

Given that there are no head-mounted displays to wear, no gloves, no cables, no huge shared screen, Park says that his new form of immersion in virtual and video space offers a personal view of cyberspace. "It's the best seat in the house," he says.

SENSE OF TOUCH

Optical technology developed to operate the first robot in space has led to commercially available controllers for computer aided design work, visual simulation applications, and to increase the enjoyment from personal computer games.

Logitech Incorporated of Fremont, California issued versions of an advanced three-dimension (3-D) controller in 1997, a device that permits users to intuitively and precisely manipulate and navigate objects through virtual worlds.

The controller has a far-reaching history that extends, literally, into space, a result of years of work by space and robotics industries. Predecessor hardware was built under contract with the Jet Propulsion Laboratory and was tasked to remotely control a robot aboard a NASA Space Shuttle/Spacelab mission in 1993. That technology has now been adapted for a wide range of tasks from mechanical design, video animation, and virtual reality design up to robotic and medical microscope control.

As an example, worldwide customers of the 3-D input device, called MAGELLAN/Spacemouse, include BMW, Chrysler, Toyota, Audi, Daimler-Benz, Porsche, and Zeiss, among many others. Engineers have found the controller particularly helpful in designing complex products, be they automobiles or airplanes. Surgeons have adopted the device to position microscopes without disturbing a surgical procedure.

A patented optical absolute measurement system imbues the 3-D controller with its impressive abilities. This opto-electronic sensor technology provides six degrees of freedom in high precision and impressive reliability.

Leading manufacturers of robots have equipped their control panels with the 3-D input tool in order to provide a powerful and reliable human/machine interface for teaching and guiding robots in the six degrees of freedom.

The controller works by providing a spring-mounted puck which the user maneuvers in order to provide motion and rotation information to the computer. In 3-D applications, the controller is used in conjunction with a 2-D mouse. The user positions an object with the 3-D controller, while working on the object using a mouse. An analogy would be a workman holding an object in his left hand and working on it with a tool held in his right hand. By eliminating the necessity of going back and forth to a computer menu, the 3-D controller increases productivity substantially in most three-dimensional applications.

Thanks to fingertip operation, the controller translates a user's sense of touch into dynamic movement of objects in those six degrees of freedom (X, Y, and Z axes, pitch, yaw, and roll).

Features of the compact controller unit include freely programmable buttons to customize a user's preference for motion control and sensitivity.

Logitech has also devised a digital game controller, an input device that lets its user move realistically in all directions. Unlike a joystick that emphasizes the physical action of a controller's hand to play a game, the digital game hardware, in a sense, connects your mind to the action. With one hand, a player controls a rubber puck that can be raised, lowered, turned, and twisted to achieve 360-degrees of movement in all directions. The other hand manages the programmable buttons.

Sophisticated combination maneuvers, such as flips and spins, can be performed using the controller without touching a computer keyboard. The speed of movements can be easily controlled, with high accuracy. Four levels of customization are offered, permitting each player the ability to create and save different configurations for each game.

Compatible with a variety of personal computer systems and entertainment software packages, the digital controller offers players the experience of being "inside" the game. With its six-degrees-of-freedom capabilities, the product is virtual reality-ready, as this new game category emerges, while offering players added realism in a broad variety of games currently on the market, notes Logitech.

By matching space program technology with a computer controller, a new dimension to game playing is attained—a winning combination where a fast mind, instead of a fast hand, is a hands-down favorite.

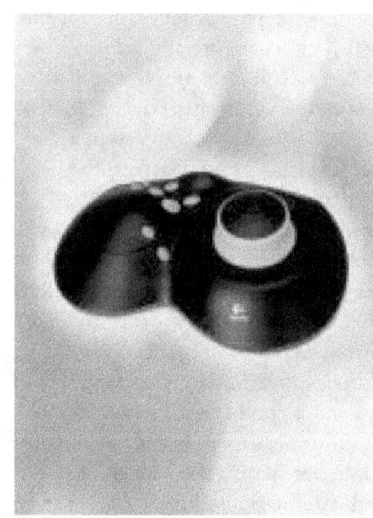

Logitech's digital game controller, based on optical technology originally developed for NASA.

JOY OF A JOYSTICK

Computer games can now be played with all the precision and sensitivity needed for a safe and soft Space Shuttle touchdown. And for good reason. A game controlling joystick for personal computer-based entertainment has been designed after the flight controllers used onboard NASA's piloted spacecraft.

ThrustMaster, Inc. of Hillsboro, Oregon introduced in 1997 an unusual type of joystick. Its design mirrors the action of the Rotational Hand Controller (RHC) built for the Johnson Space Center and Lockheed Martin by ThrustMaster. Crafted for use in Space Shuttle training simulations, the RHC is utilized by astronauts on the ground to practice runway touchdowns, as well as in orbit to sharpen maneuvering and landing techniques.

ThrustMaster produces highly realistic and innovative game controllers. The company has evolved a broad product line that includes action, adventure, sports and arcade peripherals. For some products, ThrustMaster controls have been designed after real military aircraft, giving the company an edge in several fully-functional products for computer-based flight simulators now on the market. Another market segment was met by the company with the development of auto racing controllers for driving and racing simulations.

Providing computer game enthusiasts with action on the ground and in the air, it would only seem natural that ThrustMaster would take its customers to new heights—the final frontier of space.

Lockheed Martin contracted ThrustMaster in 1996 for ten Space Shuttle control sticks and twenty electrical interface circuit boards to be used by Johnson Space Center. A specially engineered controller gleaned from the firm's Space Shuttle work is now available, featuring many new product firsts, not only for ThrustMaster, but for the gaming industry as well. The system literally brings technology from NASA's Space Shuttle program into a game player's hands. "We are creating attractive, high-quality, affordable products positioned to appeal to a more diverse user base," explains ThrustMaster CEO, Stephen Aanderud.

As a derivative of the controller built for the NASA Shuttle program, the product hosts classic features. It is touted as the next generation 3-D action game controller. Microprocessor-controlled optical sensors permit the controller's user to customize program joystick axes and buttons. The unit features four programmable axes, six programmable buttons and is completely digital. Digital technology makes the controller easy to install. Being 100 percent digital, the controller sports added advantages—smooth movement and faster response time.

The controller's most unique feature is the movement of the hand grip itself. The grip pitches forward and back from the central point of the joystick handle, allowing users to make the most of their natural hand movements for optimal game play. Along with pitch and roll control, a third axis of the joystick gives an individual the ability to yaw, emulating aircraft rudder pedals. In addition, the interactive controller features a four-way hat switch, five programmable buttons, a trigger, and a slide throttle control.

"We wanted to create the ultimate control for fast paced, 3-D game environments that would also be natural with the player's hand movements—and not require extensive learning or special training," says Robert Kubis, Product Manager for ThrustMaster, Inc. Both hands share responsibility in controlling the joystick and separate sensors immediately detect the slightest hand movements, Kubis adds.

Compatible with popular brand computers and software, the controller features a standard game port connector.

Whether there is need for immediate shields up, avoiding an errant asteroid or two, or split second shifts to light speed, this versatile action controller is sure to give any computer game aficionado a real hands-on feel for the space program.

A ThrustMaster, Inc. computer game joystick is derived from the firm's work on a Space Shuttle hand controller.

EATING ON DEMAND

There is nothing like a warm, home-cooked meal. It can be a particularly welcome delight after a hard day's work aboard a space station, hundreds of miles above Earth.

Fast-cooking technology, prompted by the requirements of space station designers at the Johnson Space Center, has led to a new generation of commercial and residential ovens. Enersyst Development Center of Dallas, Texas was engaged by NASA, as part of a space station team, to create specialized cooking equipment. The mandate was straightforward: Offer flight crews food variety, produced in a quantity and with the speed that would enable shared meals. Equally important was building a compact oven that demanded little of an astronaut's valuable time. These demands could not undermine the quality of the eating experience, nor the nutritional needs, mood and tone of space station crew members.

Enersyst tackled a number of challenges related to cooking and cooling in space. Easy preparation and operation of the device were essential; the oven had to be low-weight, yet was required to operate on little electrical energy; and the unit needed to be simple to clean and repair. Another key consideration was the control of effluents from the apparatus. Odors, water vapor, grease, and smoke produced by an oven might harm critical space station equipment. Therefore, filtration and collection control of oven fumes became more important than on Earth.

The solution from Enersyst was air impingement technology. The concept uses jets of hot air at the top and bottom of the oven that are focused on the food, rather than just heating the oven cavity as in a traditional thermal oven. By heating the food directly, foods cook faster and more consistently, retaining their flavors and textures. Embodied in Enersyst's space station prototype oven were state-of-the-art controls to provide pre-stored heating programs, all available at the push of a button.

A prototype oven for use aboard the space station has led to appliances that cook food faster, yet retain food quality.

A full manual backup system was also provided as a redundancy.

The challenges of preparing space flight cuisine were found to have a common link to an earthly reality: On-the-run consumers do not have time to cook. In many cases, people are unsure how to best prepare certain foods. Eating on demand and quality food, but in a flash, have become a part of daily life.

Enersyst has blended the air impingement idea with microwave assist, thus enhancing cooking speed while retaining the quality, taste, and appearance of the food. The microwaves heat and move water from the interior of the food to the exterior, helping in the process of cooking the food uniformly. With Enersyst technology, food can be cooked up to two to four times faster than in conventional ovens. For example, how about cooking a 12-pound turkey in just 75 minutes? Pizza, burgers, and other similar items can also be cooked in short order.

Enersyst's work on a space station oven has successfully migrated to the food service marketplace. The firm has patented the technology and has licensed it to restaurant equipment manufacturers. Vending units incorporating Enersyst innovations are being marketed in the United States, Japan, and the Pacific Rim by KRh Thermal Systems, an Enersyst licensee.

Equipment for food processing plants and restaurants, as well as vending and home applications are now part of the company's overall work in the world of food preparation. Restaurant chains like Domino's and Pizza Hut have incorporated Enersyst's air impingement technology through products offered by Lincoln Foodservice Products, Inc. and Middleby Marshall. In 1997, Enersyst began working with Thermador to establish a new line of residential ovens, bringing fast-cooking technology to the home consumer. The company's foray into the home market is one that promises to be a fast-growing venture.

Explains Sarah Palisi, Enersyst's Chief Executive Officer: "In a world where time and convenience are greatly limited, it's exciting to offer consumers the ability to enjoy the quality, flavor, and speed derived from restaurant cooking, at home."

Vending units that provide hot foods incorporate technology designed for use aboard the space station.

BACK IN TIME

A liquid-crystal tunable filter, when attached to a CCD camera, creates an imaging spectrometer, has helped in the important work of deciphering the Dead Sea Scrolls.

Cambridge Research and Instrumentation (CRI), Cambridge, Massachusetts, has become a leader in designing liquid-crystal tunable filters (LCTFs). Outfitted to Charge Coupled Device (CCD) cameras, among a host of applications, LCTFs give biomedical microscopy a new boost, providing professionals higher resolution levels and color quality that is unmatched. Partly through a Jet Propulsion Laboratory (JPL) Small Business Innovation Research (SBIR) contract, the firm developed a new class of filters under the

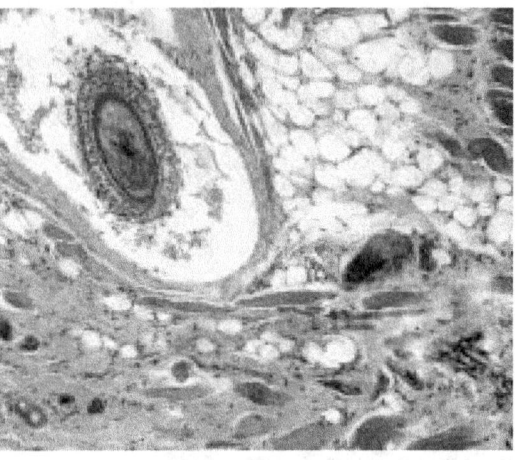

Stained rabbit skin pathology sample. Liquid crystal tunable filter technology opens up new worlds of biomedical research.

VariSpec™ product name for the construction of small, low-cost imaging spectrometers.

In simple terms, an LCTF is something like a filter wheel. But being electronic, there are no moving parts and no distortion of image between wavelengths; therefore, it is ideal for automation. Being continuously tunable, a wider range of colors is available, beyond the three fixed colors.

An imaging spectrometer acquires images of the same scene simultaneously in many contiguous spectral bands over a given spectral range. By adding wavelength to the image as a third dimension, the spectrum of any pixel in the scene can be calculated.

The saga of applying such high technology to study the Dead Sea Scrolls began with Gregory Bearman, a research scientist at JPL. He was invited to Jerusalem by the Ancient Biblical Manuscript Center to try out the LCTF-based multispectral camera on the entire Genesis Apocryphon. The specialized equipment enabled Bearman to peer at select sections of the document. The key was CRI's filter that permitted rapid switching between wavelengths. Inspecting the aged papyrus and ink, the Dead Sea Scrolls offered up a startling surprise. Once the correct wavelength was found, never-before-seen text became observable. Digital images of the documents' sections were then fed into a computer, with image-processing software further sharpening the photographs. Twenty centuries of the past were brought into crystal clarity, thanks to 20th century technology.

"Through the use of multispectral techniques, it was possible, despite the decay of twenty centuries, to read the treasure of these long-hidden writings. With this technique, the group found text never previously seen," says JPL's Bearman.

Another exciting use of CRI's LCTF has been its involvement with JPL's work on Mars rovers. The filter has successfully undergone radiation tests to qualify it for the rigors of rocket launch and jaunts to other worlds.

The assignment for Mars rovers is to scour the Martian surface, imaging the rock strewn terrain. CCD images are stacked in a computer, from the lowest wavelength to the highest, to create an "image cube." The spectrum of a selected pixel is obtained by skewing it in its third dimension, wavelength. Spectral analysis can then be performed in any of several ways. NASA is also entertaining a CRI proposal to fly an imaging instrument that can study the changing heat flow patterns at the Sun's surface.

A multitude of applications in the biomedical field alone are foreseen. Locating tumors, spotting retinal disease and doing blood chemistry are on a long list of commercial uses. In agriculture, LCTF-based imaging spectrometers are exemplary for identifying water stress in plants, as well as helping determine when a crop is ripe for the picking.

CRI sales are currently at $2 million annually and were expected to double within twelve months. Responding to rapid growth, CRI has built a new liquid crystal filter production line. Over the years, the company has won numerous awards for innovation, with demands for its hardware on the increase.

™VariSpec is a trademark of Cambridge Research and Instrumentation.

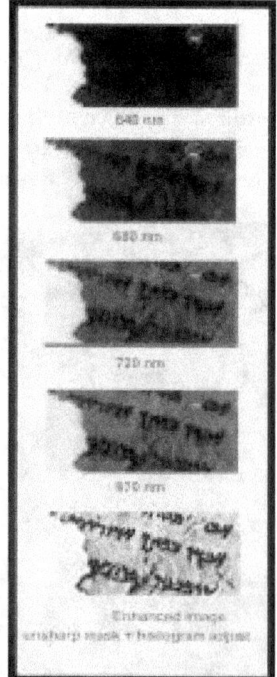

Never-before-seen writing on the Dead Sea Scrolls was uncovered by coupling liquid crystal tunable filter technology to an imaging spectrometer.

GET THE PICTURE?

Remote sensing of Earth is vital to the health and economic future of nations around the globe. Space and airborne sensors can contribute to improving agricultural products, the management of precious natural resources, and help plan and better coordinate the growth of our cities.

Positive Systems of Whitefish, Montana signed an agreement with the Stennis Space Center in Mississippi to jointly develop new technology for the remote sensing industry. This work was done under the Earth Observation Commercialization and Applications Program (EOCAP), a Stennis initiative.

Through EOCAP, improvements in Positive Systems' multispectral imaging system hauled aboard aircraft were made possible. This camera system incorporates four digital Charge Coupled Device (CCD) sensors. The sensors capture image data within discreet bands in the blue, green, red, and near infrared optical wavelengths. Coupled with use of satellite navigation, highly-accurate, real-time targeting of select ground areas is attainable. There is flexibility in aircraft observations over satellite remote sensing, particularly in addressing weather constraints to collect cloud-free imagery.

Through the agreement, Positive Systems and Stennis are tackling a problem that affects the entire remote sensing industry, including aerial photography and satellite imagery. The collaboration focuses on the company's special camera equipment that captures images in digital form, rather than on film. Digital images can be quickly turned around and easily processed by computer.

One of the company's main challenges is joining together images acquired at different times, thus having continuously changing sun angles. This creates different intensities in the images—called bidirectional reflectance—that can lead to false interpretation. New computer software is expected to solve this problem. The software integrates a complex set of correction factors, based on algorithms NASA cultivated for its own internal use.

By way of a contract, Positive Systems has adapted the NASA algorithms and is set to introduce a commercial suite of software tools for post-flight management of captured aerial images. This new generation of software efficiently prepares digital photographic images from digital camera systems, scanned film, or satellites for easier integration into advanced computerized mapping, geographical information systems and image processing applications.

NASA provided valuable assistance in the areas of characterization of the system's radiometric performance and exploration of various emerging technologies for creation of mosaic images and orthorectified imagery.

"Our clients have included urban planners, utility companies, the U.S. Forest Service and National Park

Service, defense contractors, California wine growers, and other agricultural producers," says Positive Systems Vice President, Cody Benkelman. "Through computer processing, we can clearly recognize objects on Earth that may be less than one meter in size," he adds.

Benefiting from the Stennis collaboration, Positive Systems added value to their image-taking technology and post-processing abilities. The company has completed more than 130 commercial imaging projects, and sees rapid growth in the digital aerial photography business in a number of overseas locations.

Further work on bidirectional reflectance is being made possible through an agreement brokered by the NASA-Montana State University TechLink Center in Bozeman, Montana.

TechLink, as its name implies, is linking NASA know-how to companies in Montana and surrounding states.

Aerial image showing the Cape Canaveral, Florida area. NASA launch facilities are pictured, captured by Positive Systems' airborne technology, in green, red, and near infrared bands for vegetation analysis.

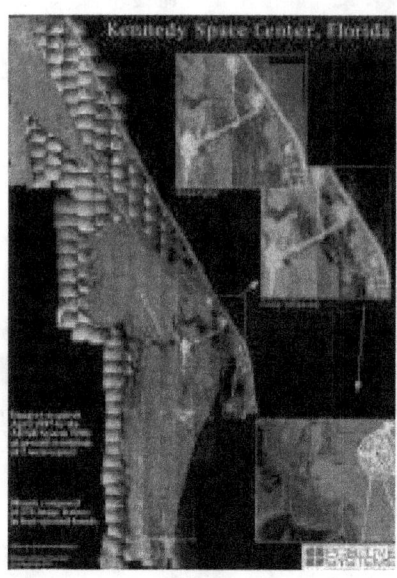

Mosaic image of Kennedy Space Center, Florida. Imagery was captured for use in ecological assessment of local vegetation. Produced by Positive Systems utilizing software from Booz-Allen & Hamilton.

CONCRETE SOLUTION

Billions of dollars worth of structures are literally eaten away by corrosion. To fight this destruction, a NASA Space Act agreement merged Kennedy Space Center (KSC) research, tied to electrical treatments of structural corrosion, with chemical processes developed by Surtreat Southeast, Inc., of Cape Canaveral, Florida.

KSC materials scientists became experts regarding an electrical treatment known as electromigration. This procedure sends corrosion-inhibiting ions to the rebar, or steel bars within a concrete slab, to prevent it from rusting, corroding and separating from the concrete.

The issue was coming up with a viable treatment that can be repeatedly applied to counter the Florida clime—a mix of salt, moisture, and baking sun. While that trio adds up to an ideal situation for tourists, it's a triple-header threat to launch pads, as well as miles of beachfront condos and other buildings.

Indeed, the aging oceanfront condominium market, as well as seaside commercial buildings, are in constant need of upkeep and restoration. The steel reinforcement bars—called rebars—that are embedded in concrete patio floors of beachside condo units have been particularly hard hit. Cracks and splitting of the concrete are major headaches. Cracks allow more water, more salt, and more acid to enter. Penetration accelerates, causing larger cracks and spalling. This troublesome material malady hits many concrete structures with decks that face the beach. Balconies in Florida have been deteriorating at a rapid pace.

With the help of Florida's Technology Research and Development Authority, an independent state agency that partners with KSC in technology transfer initiatives, Surtreat Southeast approached KSC with a chemical option to fight structural corrosion. Surtreat's GPHP product is applied to the surface of a corroding concrete slab and then seeps through to the rebar, coating it and preventing further corrosion.

"It corrects the chemical imbalance that causes the rebar to corrode. Traditional structural repair methods only last a couple of years," explains Jim Emory, president of Surtreat Southeast.

Salty sea breezes create spalling on Florida's concrete balconies. The salt migrates down to the steel reinforcing bars, rusting the bars and cracking the concrete. The edges and surfaces get unsightly and structural damage occurs in support columns.

Emory explains that concrete, due to aging and other factors, loses its pH, or acidity value. Poured concrete has a high pH value of 11, 12, or 13. That high value can inhibit corrosion. The aging of the concrete is a natural process in which the pH starts to drop. When the pH value dips into the 8 to 9 range, there is potential for corrosion of the reinforcing bars, he points out.

Combining Surtreat's GPHP with electromigration fit well in the KSC dual use program, part of NASA's technology transfer and commercialization effort. That combination is expected to result in a unique process with broad corrosion control applications. Saving money by NASA and others is anticipated by creating a structural repair method that lasts longer than just a couple of years.

The cooperative effort involved Surtreat providing to NASA the corrosion-inhibiting chemical and concrete test slabs, along with technical and staff support as needed. From KSC, testing specifications and procedures were provided. The NASA center also prepared the test slabs with the Surtreat chemical and carried out an environmental evaluation of the treatment. KSC materials scientists reviewed the applicability of the chemical treatment to the electromigration process and is preparing a report on its effectiveness following a 12-month test program.

The results could have national importance, says Rupert Lee, the NASA project engineer leading the joint effort. "Any breakthrough in corrosion mitigation technology will have a significant impact on the integrity of this nation's infrastructure," he explains.

The cracked (spalled) concrete must be chipped away and replaced. The Surtreat chemical is sprayed on the concrete and exposed steel bars.

THE INSIDE VIEW

Technology developed for the Apollo lunar landing program has come full circle—launched to the moon and then finding a home back here on Earth—advancing the art of computer imaging inspection.

Computed tomography, widely known as CT or CATScan, is a medical diagnostic technique for comprehensive body scanning. It incorporates digital image processing technology that traces its origin to NASA research and development performed as a prelude to the Apollo program. Millions of people around the world benefit each year from the medical applications of this technology.

This same capability has found secondary utility as an aerospace, and general industrial tool—the Advanced Computed Tomography Inspection System, or ACTIS. In its medical role, CT scans the human body for tumors or other irregularities. The ACTIS system, on the other hand, finds defects in structures and components, be they castings, assemblies, rocket engines, or nozzles. The system helped NASA's Space Shuttle engineers characterize structural assemblies by producing high-quality CT images. These images demonstrated the ablative properties of various solid rocket motor nozzle assembly materials, revealing anomalies at bondline interfaces. Flaws such as these could cascade into mission failures if not detected.

For Marshall Space Flight Center (MSFC), Bio-Imaging Research, Inc. (BIR), Lincolnshire, Illinois, developed an industrial inspection system that employs CT technology for nondestructive evaluation. The system, designed and built by BIR under contract to NASA, was installed in 1988. Two years later, Boeing Aerospace in Seattle, Washington purchased an ACTIS system for nondestructive CT inspection of a variety of commercial and military aerospace components.

"ACTIS has upheld NASA's reputation for technology that benefits industry and science," advises Richard Bernardi, BIR's Vice President for Business Development. It was used to view the interior of a sealed centennial "time capsule"—an accomplishment that was featured in several national publications such as *National Geographic*.

Cross-sectional CT images offer more detail than radiographic images. The equipment's high-speed scanning feature offers the capability for 100 percent inspection in a production environment.

Looking back in time, ACTIS was used to examine the skull and caudal vertebrae of a Tyrannosaurus Rex while the fossils were still imbedded inside their rock matrix. The images helped paleontologists determine the best way to recover the valuable fossils undamaged.

More recently, yet another successful ACTIS spinoff took place. The NASA MSFC ACTIS system proved the feasibility of CT inspection of 55 gallon drums of radioactive waste for the U.S. Department of Energy (DOE), based on tests with surrogate (cold) drums. There are more than one million drums of waste dating back to the beginning of the atomic era. Located in interim storage, these drums are subject to corrosion and potential leakage. They cannot be transported across state lines or permanently disposed of until they have been examined (characterized) to prove that they neither contain more than one-half percent free liquid, nor that the drum wall has lost integrity.

Identifying these conditions is ideal for computed tomography. According to Bernardi, preliminary tests on cold surrogate drums at MSFC were so successful, DOE has since invested in a $6 million BIR program to develop and field test a transportable (mobile semi-trailer), two-million volt x-ray CT scanner. That hardware uses many of the design principles and elements similar to the BIR ACTIS CT scanner currently at NASA's Marshall Space Flight Center. The mobile drum inspection unit uses a combination of x-ray and gamma-ray technology to "fingerprint" the source of radioactivity. Characterization is essential to decision making on how to treat or transport the waste.

Industrial inspection system, built by Bio-Imaging Research, is ideal for nondestructive testing of aerospace structures and components. The technology is rooted in NASA's Apollo moon landing project.

Testing of this new technology application that fingerprints nuclear waste drums is also underway at several DOE sites. Labeled the Waste Inspection Tomography (WIT) system, the equipment promises to revolutionize the nuclear waste drum characterization process.

Computed tomography, widely used in the medical field, has proven invaluable to locate imperfections in aerospace components. This Advanced Computed Tomography Inspection System can scan anything from small turbine blades, auto parts, and drums of nuclear waste to rocket assemblies.

EVERY BREATH YOU TAKE

Sustaining astronaut crews on lengthy tours-of-duty demands the utmost in air revitalization hardware. In fact, their very lives depend on quality air. Earth-orbiting space stations, piloted spacecraft en route to Mars, or habitats deployed on the Moon require a controlled ecological life support system (CELSS). This type of apparatus relies on the growth of healthy plants and the removal of any contaminants expelled in the process.

KSE, Inc., located in Amherst, Massachusetts, is building efficient air revitalization systems for space missions. The work is underway through Small Business Innovation Research (SBIR) contracts with Johnson Space Center. KSE is preparing a photocatalytic ethylene scrubber for the constant removal of ethylene from space-rated plant growth systems. Maintenance of ethylene levels in the parts-per-billion range in plant growth facilities is essential for a CELSS, which relies on the growth of healthy plants.

The KSE technology is identified as Adsorption-Integrated-Reaction Process, or AIR Process for short. For NASA's interests, the company adapted the process to develop the Vehicular Air Purification and Odor Reduction System (VAPOR).

This technological innovation represents a novel integration of separation and purification unit operations. Unique photocatalyst formulations are exploited. A proprietary catalyst traps and concentrates contaminants onto the surface of a highly effective adsorbent. This process then destroys the contaminants at ambient temperatures in the presence of ultraviolet light. That promotes dissolution of the contaminant into environmentally safe compounds, such as water and carbon dioxide. The photocatalyst's surface is restored to capture more impurities.

KSE has prepared and tested over 100 photocatalysts in the program, in addition to hundreds of previous photocatalysts. Plans to fly plant chamber test units in 1998 are under discussion.

Extensive applications in commercial pollution control are also anticipated for this novel, cost-effective technology. Those applications include air quality and remediation of soils and groundwater.

Air streams carrying volatile organic compounds are often found at hazardous waste sites undergoing remediation.

KSE air purification hardware was developed under contract to NASA for removal of contaminants in closed life support systems.

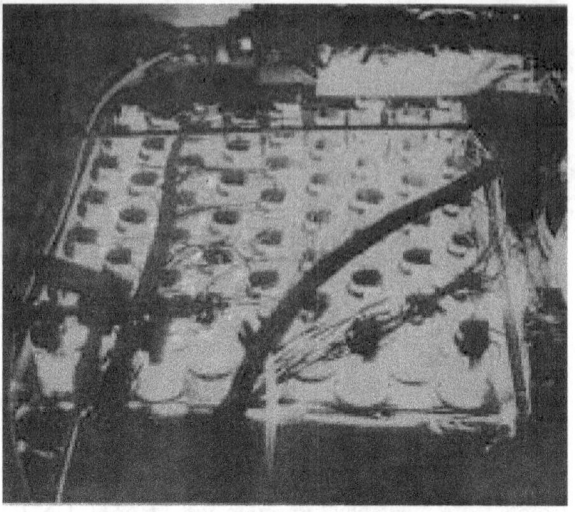

Contaminants are trapped on a novel catalytic adsorbent. In the presence of ultraviolet light, the trapped contaminants are catalytically destroyed.

For example, groundwater stripping, or soil vapor extraction, produces large volumes of air containing the volatile organic compounds that must be removed. In operation, KSE's process cleans the effluent air, yielding carbon dioxide and water, which are carried out in the air stream exiting the reactor. For chlorinated volatile organic compounds, the chlorine atoms are converted to hydrogen chloride with some chlorine gas. If needed, these can be removed from the air stream using conventional scrubbers or adsorbents.

KSE also has designed process reactors that churn out no residual wastes or by-products that need further treatment or disposal as hazardous waste. The process is self-contained and mobile, requiring a small amount of space. Moreover, this equipment needs less energy than thermal incineration or catalytic oxidation. In addition, it has lower total system costs than traditional technologies. Having flip-of-the-switch, on/off controls, the adsorption-integrated-reaction process has shown up to 99-plus percent destruction efficiency in field tests.

A commercial customer has purchased from KSE a full-scale air purification system. The reactor can destroy airborne trichloroethylene, dichloroethylene, and vinyl chloride vapors. Those chemicals are resident at an Environmental Protection Agency Superfund site. KSE's photocatalytic approach was seen as the most cost-effective technology by an independent consulting firm. The firm's reactor system was judged far superior to activated carbon adsorption, catalytic oxidation, and thermal oxidation.

Another purchase order is being discussed—a unit eyed for installation at a municipal Superfund site in Vermont. If placed at that site, the KSE technology will retire an existing carbon adsorber unit.

The technology received the 1997 SBIR of the Year Award for Environment, Energy & Resource Management by *NASA Tech Briefs* and the Technology Utilization Foundation.

NO MOVING PARTS

Shaving off weight on a spacecraft translates into savings on launch costs. There is an increasing demand for satellites that are cheaper, better, and faster to produce—all factors in space technology that give more service for the dollar spent.

Research International, Woodinville, Washington, has created a solid-state micromachined pump for cooling electronics in space. The work was made possible through NASA Small Business Innovation Research (SBIR) contracts, awarded to the firm from Johnson Space Center and the Lewis Research Center. Other functions for the pump include circulating heat transfer fluids, and monitoring fire and gas hazards aboard naval warships. In the NASA-sponsored work, pumping action was based on the interaction of certain fluids with impressed electric fields.

Programs that demand micromachining involve a number of factors: Complex fluid flow structures etched in refractory glass; multilayer films deposited to provide corrosion resistance over years of operation; thin fixtures made from epitaxial silicon using etch-stop techniques; and the perfection of hermetic sealing techniques.

Micromachining is but one focus area of Research International. Significant in-house capabilities are in place for custom instrumentation, optical sensing, and thin-film development.

Commercial applications for these technology thrusts are many. They include detection of toxins and pollutants in coal mines, as well as an early warning smoke detector for industrial applications. The prospects for using fiber optic probes to determine the air content of freshly-poured concrete are also under review by the group.

Research International has incorporated the micromachined, no-moving-parts pump into a four-channel, solid-state fluorometer. A patent is pending for this product, MANTIS™, a tightly packaged, portable, fully-automated immunoassay system.

There is currently a need for new technologies that are designed specifically for the high-sensitivity field monitoring of toxins, explosives, and chemical contaminants. Some of the most promising strategies for carrying out such strategies are based on studying the immune and protective responses of animals and humans.

The solid-state fluorometer system integrates optics, electronics, and software into an all-in-one way to monitor the progress of immunological reactions. Using this system, toxins such as *Y. Pestis* have been detected at levels below one part per billion from samples of a few hundred microliters.

Research International's micromachining methods are being used for the construction of miniature fluidic devices for use in medical drug delivery. The high-tech company is on contract with NASA to develop a rapidly responding fiber optic-based hydrogen gas sensor. This is based on work performed on interferometric devices that reversibly change optical properties via an internal chemical reaction.

In another NASA program, Research International has developed a novel optically-based compact particulate monitor that can be used for the detection of smoke or other airborne matter in the space shuttle.

^{TM}MANTIS is a trademark of Research International.

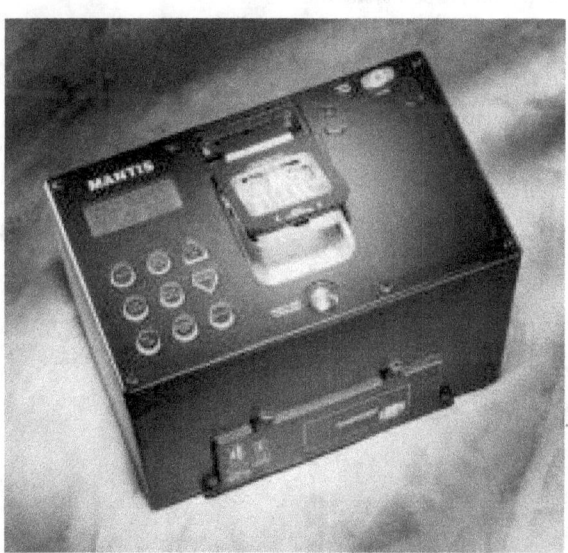

NASA-supported research helped in the creation of micromachined pumps. This appliance is integral to a number of Research International products, like this highly-sensitive portable unit that detects toxins, explosives, and chemical contaminants.

ENTREPRENEURIAL ASPIRATIONS

"Intelligently Interactive Web Sites" is a marketing declaration of MiraNet, Incorporated, a Silicon Valley startup with proprietary software expanded from an expert system long used at NASA.

MiraNet of Palo Alto, California is one of two companies that have spun off from Recom Technologies of Roseville, California. The other firm, Attention Control Systems, Incorporated of Mountain View, California is utilizing intelligent planning software that Recom developed for the Ames Research Center's (ARC's) Computational Sciences Division.

In the fast-paced, head-spinning, computer technology-driven world of Silicon Valley, there is one obvious linkage between the three high-tech entrepreneurial organizations—NASA technology. But first, a little history.

Recom Technologies was established in 1980 by Jack Lee, a 15-year veteran of ARC involvement, working as a contractor employee. Areas of expertise for Lee centered on simulation and modeling, information systems development, engineering applications, artificial intelligence, and expert systems development.

"Entrepreneurial aspirations," got the best of Lee, he says: The ability to pursue those aspirations, particularly as they related to creating products and services resulting from scientific research. After forming Recom Technologies, Lee later was awarded an ARC contract to assist the NASA center in producing software to control a telescope mount. Just a year passed, with Recom staffed with scientists capable in the area of expert systems, software engineering and simulation systems.

Recom Technologies has grown quickly, with the company's nearly 450 employees proposing a broad range of business-oriented research topics, from fiber optic sensors that recognize brain tumors to a virtual reality training system for surgical procedures and next generation video games.

Eventually, two efforts were chosen, along with two subsidiary companies formed to permit ownership by the employees who generated the concepts. Attention Control Systems focused its energies on a hand-held device, employing real-time planning software together with neuropsychology therapy concepts. This device is being prepared as an aid in cognitive rehabilitation of brain injury patients. It can also assist other people with psychological disorders such as Alzheimer's Disease and Attention Deficit Disorder. The core system makes use of intelligent planning software that Recom developed for ARC. The hardware provides patients with detailed daily activity scripts with the help of their therapists and provides a major step into autonomous living.

Recom has committed to spinoff an average of one subsidiary per year.

Sound a little ambitious? As Recom's Jack Lee is fond of saying: "Anyone can provide technology...Recom's people deliver solutions."

MiraNet was founded to bring NASA technology to the Internet and the Web, and its first product, WEXpert™, is a web-enabled version of the NASA-developed, rules-based expert system called CLIPS.

WEXpert guides web visitors to answers in an intelligently interactive process. Although the system appears to be a web site, there are only the WEXpert engine plus a knowledge base in the form of text-based rules. Web pages are automatically generated for each visitor based on the answers given during a consulting session. No technical programming is needed to set up a WEXpert Guided Web, only English-like rules that reflect the type of knowledge or expertise being dispensed.

Where typical web sites are a jungle of information requiring perseverance, search engines, and Internet surfing skills, WEXpert guides you to answers or advice. It is targeting corporations who wish to enhance customer service or provide web-based sales advice. The growing frustrations of telephone menus and waiting for human support are turning a growing population of people to web sites for help. WEXpert will make web sites friendlier and more productive, says Alex Cheng, President of MiraNet.

WEXauthor™
This figure depicts a sample knowledge base used to diagnose problems in photography. By defining important attributes and creating rules based on these attributes, the knowledge base designer maps the path to a correct diagnosis.

™ WEXauthor is a trademark of MiraNet.

™ WEXpert is a trademark of MiraNet.

A VIRTUAL WORLD OF VISUALIZATION

Under the hood and through the furnace.

Those are two places where industry has gained cost-saving advances from spinoffs of NASA aerospace computer technology. That is the way that M. Gene Konopik sees it as President, Federal Systems Group of Sterling Software, Incorporated in McLean, Virginia. Sterling Software has had a 25-year history with NASA.

Advanced three-dimensional (3-D) interactive graphics were pioneered at the Ames Research Center (ARC). These same tools have been adapted and adopted in ways that now help advance automotive virtual reality models, heat flow in furnaces, and air pollution.

Starting in the early 1980s, ARC made significant investments in pioneering research and development for scientific visualization. ARC developed 3-D graphical data formats and basic graphics codes for displaying computational fluid dynamics (CFD) results.

Sterling Software began assisting NASA by building the first workstation software packages for 3-D scientific animation. While these early packages were originally built for aerospace CFD graphics, prospects began emerging for many other applications in visualizing similar 3-D grid-oriented data.

In 1990, Sterling Software developed a toolkit for ARC on contract. As a workstation-based modular analysis and visualization tool, animated grids and grid-oriented data can be derived. While constructed in modules, each module of the tool operates as an independent process. They are under control of a central process that maintains the toolkit's data in shared memory. A few of the current modules include: interactive surfaces, vectors and contours; generation of isosurfaces and arbitrary cutting planes; unstructured data analysis; grid quality, resolution, and geometry inspection; and computation of scalar and vector CFD functions and custom algorithms.

Sterling Software's visualization toolkit is most widely used for reviewing fluid flow and similar types of grid-oriented 3-D data. The Space Shuttle, jet engine turbine internal flows, vertical short-takeoff-and-landing ground effect research, and the vortex dynamics of whirling helicopter blades—these are examples of technologies where the visualization software has proven of great merit.

On a non-exclusive basis, Sterling has created special-purpose versions of its software toolkit. Designed with industry and non-NASA customers in mind, the visualization software can now be run on a wider set of workstations. Ford Motor Company, for example, has made use of Sterling's toolkit and modules to visualize under-hood and under-body air flow and heat build-up. Sterling engineered a special package of virtual reality-based computer software to help Ford shape interactive virtual work sessions. The automaker is investigating ways engineering and design personnel—either at the same

spot or distant locations—can dynamically adjust a mathematical car model, observing the impact of those alterations in real-time or near real-time.

For the U.S. Environmental Protection Agency, Sterling adapted software versions to visualize acid rain deposition and ozone depletion. It is clear that global climate change research is likely to push the frontiers of data visualization technology to a great degree as the globe prepares to enter the 21st century.

Sterling Software's long-standing relationship with Ames Research Center continues in 1998. Recent contracts with NASA support the company's involvement in the NASA Science Internet at ARC. Serving over 25,000 worldwide users engaged in NASA-related science and research, as well as educational collaboration, ARC's NASA Science Internet was created and is maintained by engineers from Sterling Software.

Three-dimensional visualization was stimulated by NASA in the early 1980s for scientific data display purposes.

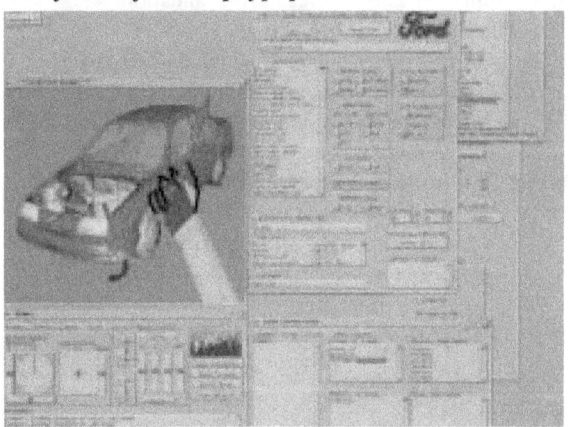

Sterling Software's visualization technology has made a mark on the automobile industry by allowing under-the-hood changes in a virtual reality simulation.

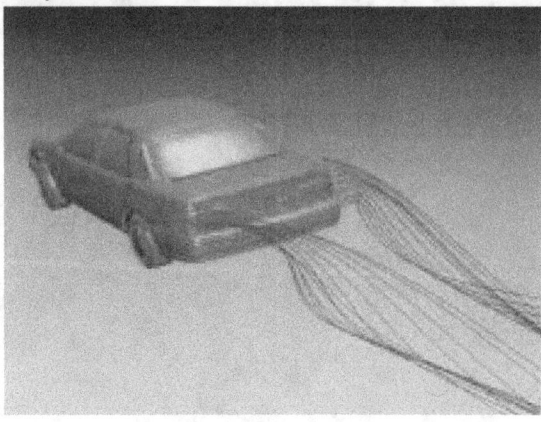

MINING FOR DATA

How can you convert data into knowledge needed to make better predictions and decisions?

AbTech Corporation of Charlottesville, Virginia has developed advanced system-level diagnostics software. The firm's ModelQuest™ System Validator software was initially created for the Dryden Flight Research Center (DFRC) under a Small Business Innovation Research (SBIR) grant. The software was written to assist system engineers in validating an automated research flight control system for the F-18 High Alpha Research Vehicle (HARV). Barron Associates modified NASA's F-18 HARV simulation software for this project.

In the commercial world, there are many types of control systems. Costs can run into the millions for validation of complex systems. For example, Boeing, Lockheed Martin, and other industry giants spend between one and seven person-hours per line of code on the validation process. Although some costs are reduced by streamlining the process, a streamlined process does not eliminate the need for extensive validation.

The diagnostics software is applicable to aircraft flight control systems, engines, manufacturing processes, medical signal processing, and mechanical equipment for:

- Validation of new and modified systems and simulations;
- System-level diagnostics (anomaly detection, isolation, prediction);
- Comparative signal analysis (satellite telemetry, financial time-series and medical diagnostics).

In simple terms, the diagnostic software reduces the time and cost of validation and diagnosis of complex analog systems, increases confidence in the validity of the systems, and automates the development of superior system-level diagnostic mathematical models. It applies the mathematical-modeling techniques to "overplots" of hundreds of simulated anomalous signals with "truth" signals to learn automatically the differences between normal and anomalous systems.

Given recorded signals from a system being evaluated, the resulting software overplot analysis models identify the probability that a design anomaly or failure exists, estimate the level of degradation of performance, and attempt to isolate the cause of anomaly or failure.

The highly automated capability of this software can yield a striking reduction in the time needed to validate critical systems or substantially increase the level of fault detection and isolation.

Successful applications include decision support that recommends treatments for grasshopper infestation. In the area of healthcare, a savings of 60 percent was achieved in a commercial hearing test product. As a stock market prediction application, AbTech's data mining tools have proven the fastest and easiest to use over six different neural network products—and provided by far the best results.

"Improvements in information technology are motivating business and technical professionals to rethink their strategies to discover better information to compete more effectively," comments Gerard Montgomery, AbTech's President and Chief Executive Officer. "AbTech is poised to capitalize on this rapidly growing industry by delivering superior data mining tools to companies that realize the need to analyze and understand their data," he concludes.

In its sixth consecutive year as a profitable data mining software company, AbTech has grown since 1992 to have a customer base of over 4,000.

™ ModelQuest is a trademark of AbTech Corporation.

Data mining software tools empower users with decision and prediction skills. The tool can be applied to development of flight controls for high performance aircraft.

GRID WORK

An interactive, graphically-oriented software system that generates three-dimensional, multiple block, structured grids has gained wide popularity with scientists and engineers.

Originally developed for the U.S. Air Force by a major aerospace firm, the software produces computational grids for engineering analysis, primarily computational fluid dynamics (CFD). Both Langley Research Center and Ames Research Center played important roles in expanding its use throughout NASA, to solve more complex CFD problems with greater efficiency.

NASA delivered the software in 1994 to the Computer Software Management and Information Center (COSMIC). Government-developed computer programs adaptable to secondary use are maintained at COSMIC. That same year, the software's source code was made available for enhancement and commercialization. Today, Pointwise, Inc. of Fort Worth, Texas is offering for commercial purchase the augmented grid software.

"NASA's willingness to distribute this technology allowed more than 300 individuals at various governmental and industrial agencies to gain quick access to the software," says John Steinbrenner, Pointwise Vice President, Research & Development.

Pointwise is a supplier to the engineering analysis community of the grid generation software known as Gridgen. A complete line of services is centered around the software. Scientists and engineers prepare geometry data for computer simulations of fluid flow by tapping the strengths of the software system. The enabling software bridges computer-aided design (CAD) software and analysis software such as CFD and finite element analysis (FEA). A favored tool in the demanding aerospace industry, the high quality grids are in use in fields ranging from automotive to turbomachinery.

A unique aspect of Pointwise's grid software is that it decomposes an experimental shape's three-dimensional domain into contiguous sub-domains called blocks. Grids on the surface of each block are generated, followed by the volume grids within.

"It's like a computer-aided design system," says Pointwise's president, John Chawner, "where instead of drawing the experimental shape, we're creating the grid around it. Our approach to making complex problems as simple as possible is to make the code graphical and handle the bookkeeping automatically."

Challenging shapes? No problem, as multiple block, structured grids, and three-dimensional space can break into any number of contiguous blocks, each in its own computational space. Furthermore, the system defies memory limitations. A flowfield, for example, that demands millions of grid points may exceed available memory. But, by apportioning grid points among several blocks and doing the calculations on them one at a time, the flowfield can be resolved.

Speed...space...and time. The software can render the flowfields slipping over the supersonic skin of a high-performance jet aircraft. This can be done as efficiently as laying out the surface grids for the bifurcation region of the carotid artery.

Pointwise's flagship software has been distributed to over 300 sites and has been used to create grids for spacecraft, fighter aircraft, internal and external flowfields, missiles, submarines, surface vessels, sailboats, and turbomachinery. A wide divergent group of users now depends on the powerful software system, be they in industry, government laboratories, or academia. The company has formed alliances with two other firms, further expanding the development and sale of the software to aerospace, automotive, and mechanical engineering markets.

Since it began selling the software in 1995, Pointwise has released refined versions of the product. In July 1997, the company announced the availability of version 12 of the numerical grid generation software. In 1998, they will introduce an even more powerful version that includes the ability to create unstructured grids for handling even more complex shapes.

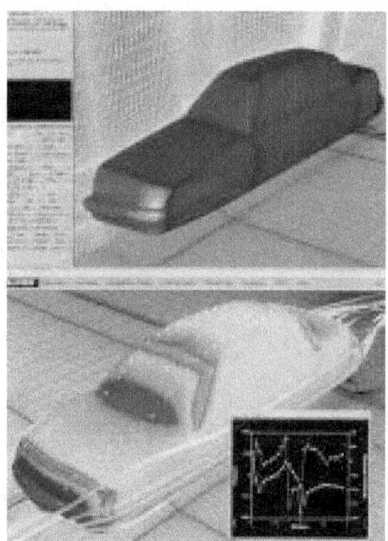

Pointwise, Inc. has updated and modified government-produced software for commercial sale. Computational grids produced by the software are of great help in automotive design work.

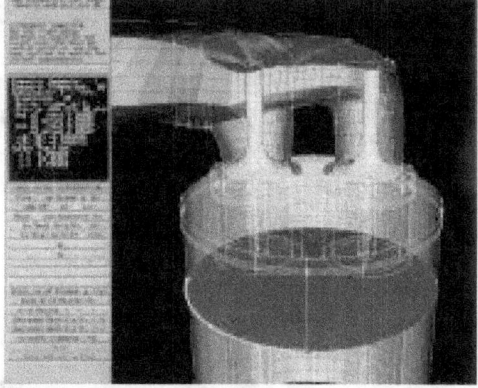

Computer-generated grids for engineering analysis have become a boon for manufacturers. Software has been made available from NASA's COSMIC system for secondary use of government-developed computer programs.

TAKING THE HEAT OFF THERMAL LOADS

A turn-of-the-century telescope is under development at Goddard Space Flight Center. The Next Generation Space Telescope (NGST) is a follow-on to the already orbiting Hubble Space Telescope. NGST will be imbued with the ability to visit a time when galaxies were young. This pioneering, highly advanced scientific observatory will observe the "dark zone," a period when primordial seeds began to evolve into the galaxies and stars we see today.

A major engineering challenge awaits NGST builders. A tool that is being employed to help engineers grapple with the complexity of the NGST is offered by Frederick A. Costello, Incorporated (FAC, Inc.), a concurrent engineering software developer based in Herndon, Virginia. Costello's TCON™ software enables graphical user-interface programs to create three-dimensional thermal models. The NGST is to be passively cooled. Solar heating must be reduced by a factor of 10,000 in NGST. The observatory's proposed capabilities will far exceed anything possible from the ground or in space.

Costello's software is an affordable approach to create full three-dimensional thermal models efficiently and accurately. The product reduces modeling time to one-third the amount of time required by conventional, less-automated methods. TCON software can convert over 4500 nodes and 4500 radiation surfaces. Color temperature contours make presentation of the software's assessment of thermal characteristics quick, easy, and effective.

Goddard Space Flight Center awarded Costello a Small Business Innovation Research (SBIR) contract to develop a low-cost computer program for integrating thermal modeling with other engineering disciplines. That SBIR funding sparked the innovative software, developing a whole new line of business for the consulting company.

The software has been sold to NASA and to major American and foreign companies. It has also been adopted by developers of commercial geosynchronous Earth orbiting (GEO) and low Earth orbiting (LEO) satellites.

Since 1978, FAC, Inc. has offered engineering design analysis and testing services. The company has served a range of clients in the government—about 20 percent of the business—with the remaining 80 percent representing the private sector.

FAC, Inc. engineering modules can also be used to define and view the orbit of a spacecraft interactively on the screen. Attitude control modes, attitudes, and altitudes can be adjusted while viewing the orbit, as environmental flux data are updated on the screen for relative positioning of the Sun and the Earth with respect to each axis of the spacecraft. Defining spacecraft attitudes for a spinning or trajectory-bound spacecraft is automatic, no matter what the spin axis. Another software module provides the engineer with an automated method of transferring the temperature data from the thermal analyzer to the structural analyzer, even when the two models are dissimilar.

Future versions of the software are being readied. These modifications permit software interfaces with other graphical user-interface programs and thermal analyzers.

In the field of aerospace thermal design, FAC, Inc. has worked from cryogenics to ablation. Industries served by the company include electronics firms, aerospace companies, makers of industrial processes and equipment, energy technology groups, and the defense market.

FAC, Inc. has designed many temperature-control systems for spacecraft and instruments, such as the Hubble Space Telescope, the TOMS Earth-monitoring probe, and the Compton Gamma Ray Observatory.

™TCON, Rapid Thermal Modeler, OrbitPlotter, and MAPBACK are trademarks of Frederick A. Costello, Incorporated.

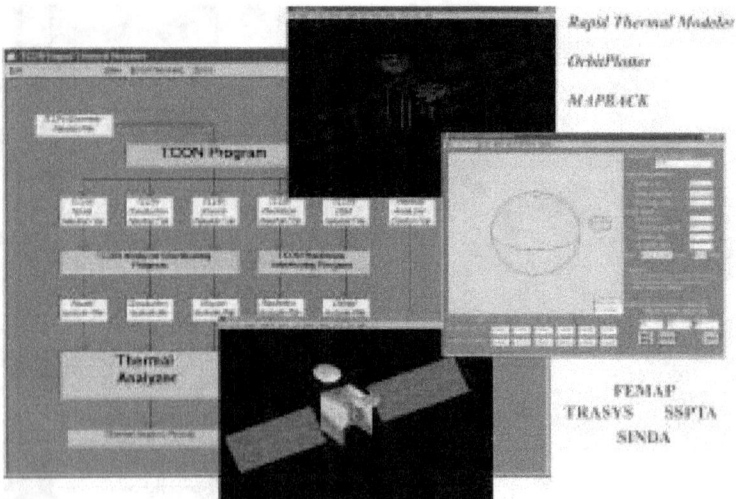

TCON™ Concurrent Engineering Software, is a thermal translation program that enables graphical user-interface programs to create full thermal models efficiently and accurately. The different modules of TCON are the Rapid Thermal Modeler™, OrbitPlotter™, MAPBACK™, and SINDAFAC.

GO WITH THE FLOW

A way to improve the process of applying a touch of thin-film to materials has emerged from weighty research in microgravity.

Nektonics, Inc. of Cambridge, Massachusetts, is dedicated to computational fluid dynamics (CFD), including research and development in CFD software and application of CFD techniques to engineering analysis and design. The company specializes in medical device design and coatings applications.

Started in 1987 with students graduating from the Massachusetts Institute of Technology's fluid mechanics laboratory, the firm has been awarded Small Business Innovation Research (SBIR) funding from the Lewis Research Center. Resulting from the SBIR work, Nektonics designed a powerful simulation software, called NEKTON, for the modeling and analysis of a wide range of coating flows. The software was subsequently used in the Surface Tension Driven Convection Experiment (STDCE), a low gravity fluid physics investigation flown on two Space Shuttle missions, the June 1992 flight of STS-50 and aboard STS-73 in October 1995.

Without the strong gravitational effects which normally dominate fluid flow behavior on Earth, surface tension forces primarily determine the fluid behavior in space.

This phenomenon has been graphically illustrated by astronauts drinking through a straw, floating balls of orange juice in the microgravity environment. On Earth, the same surface forces dominate the behavior of fluids when the size of the fluids is small. For example, the surface forces cause very small amounts of water to coalesce into spherical droplets. In another example, thin-film coating processes are largely governed by surface tension effects.

Nektonics researchers have identified commercial down-to-Earth uses of their space-related CFD research. The work done for Lewis Research Center has been adapted as a flagship software product for Nektonics. That tool is now used in a broader market by many companies to analyze and improve thin-film coating processes.

Companies utilizing the simulation software can determine how surface tension determines the thickness and affects the quality of a coating.

Several advancements to the initial CFD software product have been made by Nektonics for industrial customers. Analysis of a wide range of coating flows including thin-film coating analysis, polymer processing, magnetic media, and glass melt flows is possible using the coating process simulation software. Polaroid, Xerox, 3M, Dow Corning, Mead Paper, BASF, Mitsubishi, Chugai, and Dupont Imaging Systems are a few of the companies that presently use the Nektonics product.

In another commercial use of the NEKTON simulation software, many years of testing were saved in the fabrication of the first continuous glucose monitor. Using the CFD simulation software, Cygnus, Incorporated designed the monitor that measures glucose levels through skin contact, eliminating the need for diabetic patients to prick their fingers to obtain blood. The ability to evaluate different designs—making use of Nektonics technology—without building prototypes, drastically reduced the amount of time needed to bring the glucose monitor to market.

Armstrong World Industries used the CFD product from Nektonics to predict the scale-up of a pilot plant air knife coating operation.

Nektonics products have found utility in paper, photographic, imaging, and printing industries for the analysis of thin coating flows. In particular, the product is ideally suited for displaying stream lines, thereby showing how the yield stress in the liquid affects the flow field and creates stagnate areas.

Nektonics is in the process of adding several new features and enhancements to its software product. The most important addition is the ability to analyze the stability of steady two-dimensional calculations to three-dimensional perturbations. This improved version of the firm's computer code is expected to be released in mid-1998.

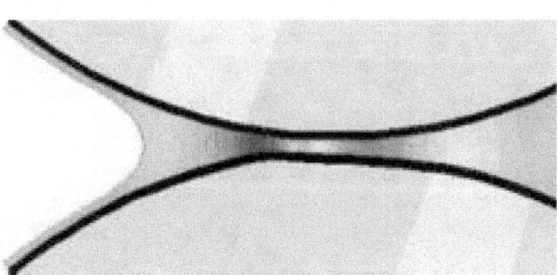

In many roll coating applications, one of the rolls may be rubber-backed. As shown in this pressure distribution in a forward roll coater, NEKTON can predict the deformation of the soft rubber cover and its effect on flow split and meniscus location.

NEKTON's deforming mesh technique makes for easy setup and solution of problems involving unknown free surface locations, even those involving more than one fluid as shown in this two-layer slot coater application. In this case, the location and shape of the interface between the immiscible fluid layers is also computed.

PIXEL PARADISE

Everyone has been impressed by those sharp images relayed to Earth over billions of miles, courtesy of outbound spacecraft as they reach other worlds. Each picture element of an image sent back from spacecraft is known as a pixel.

If PixelVision, Beaverton, Oregon, has its way, you haven't seen anything yet! Scientists at the Jet Propulsion Laboratory (JPL) and PixelVision engineers have developed active-pixel technology with noise performance on a par with Charge Coupled Device (CCD) detectors. While CCDs provide impressive sensitivity, dynamic range, and linear response, readout speeds of CCD images have lagged. That fact has imposed performance limitations on large-array cameras.

JPL researchers, aided by PixelVision, had a challenge on their hands. A NASA mission to faraway Pluto is on the books. The long journey will surely tax technology to the max, including the need for a high-sensitivity Pluto imager.

The answer came through development of an active-pixel sensor that uses very large scale integration, incorporating all necessary camera functions onto a single chip. Each pixel contains not only the photodetector element, but also active transistor circuitry for reading the pixel signal. Unlike CCDs, which need to add circuitry to prevent light distortion, the active-pixel sensor is intrinsically not prone to electrical charge leaking from one pixel to another when exposed to intense light.

JPL's goal for Pluto was a video-rate read noise of less than five electrons, a level of performance previously available only in slow-scan scientific cameras. The end result from JPL engineers and PixelVision researchers was a high-speed, video-rate CCD camera that retains resolution and has an impressive readout speed of 30 million pixels per second. The instrument has been labeled a high-speed imaging engine.

Fabrication of the Pluto CCDs required submicron precision across the entire semiconductor wafer, several inches in diameter. That requirement pushed semiconductor processing technology close to its current limits. State-of-the-art technology was also needed to avoid image smearing, given the high readout speed of the camera.

Subsequent to the development of the high-speed imaging engine for Pluto, PixelVision has begun to address various applications of the technology.

According to George Williams, Vice President and General Manger of PixelVision, the high-speed imaging engine is ideal for applications that require video-rate imaging for real-time monitoring of dynamic events. One example, Williams says, is fluorescence imaging of energetic events in cell biology. Still other uses for the technology fall into enhanced night vision for military and surveillance purposes. The Pluto imager yields a robust night vision capability. Moreover, it does so at far lower power levels than vacuum tube designs that are part of airborne visible night vision systems in wide use today.

The imager is also well suited to applications that require high speed simply to handle a large volume of image data, such as three-dimensional reconstruction microscopy and film digitization. Increased CCD speed and ease of integration will both be important benefits to manufacturers and end users of reconstruction microscopes, Williams says.

CCD camera offers sensitivity of slow-scan scientific camera with extremely fast readout. Built for a mission to Pluto, the device supports several commercial applications, such as night vision surveillance.

SIMULATION ACCELERATOR

There is performing...then there is outperforming.

In this category is EAI Simulation Associates, Inc. of West Long Beach, New Jersey, a design leader in supplying parallel systems for real-time high fidelity simulation.

Under Goddard Space Flight Center Small Business Innovation Research (SBIR) contracts, EAI Simulation Associates developed a new digital simulation computer called the Advanced Real-Time (ARTS) System. Its architecture is based on the analog model of computation instead of the classical von Neumann model. SBIR work included development of a simulation product that outperforms all other computers, including supercomputers, on a wide range of continuous system simulation applications. The successful completion of that NASA work led to the formation of the company in 1994.

EAI Simulation Associates has a joint marketing, distribution, and maintenance agreement with Halifax Corporation for the commercial sale of new all-digital simulation products. Among them is the SBIR-funded computer product, a data flow processor that provides supercomputer performance. Five ARTS systems have been sold including over $2 million in sales to Japan. The product has been used in numbers of applications to support aerospace and automotive work, as well as in the electric power and chemical reactor industry.

The truly innovative feature of the accelerator processor resides within the scheduling compiler. This compiler does not translate code into an intermediate language such as C or FORTRAN. Instead, it breaks down source statements into threads of elementary operations. These operations are then automatically scheduled to execute directly, with maximum fine-grain parallelism, on very long instruction word hardware.

Marketing of the data flow processor by Halifax is underway, a product-add to that company's line of flight simulation systems and computer simulation services to civilian and military users.

The accelerator designed by EAI Simulation Associates is the only digital system that offers never-before-attained computational speeds. Put to the test, a real-time simulation of a Space Shuttle main engine was undertaken. Stable and accurate results were achieved, with the system simulating Space Shuttle main engine controllers, high-speed turbo pumps, and other elements. Temperatures, pressures, and flow rates were computed for fuel, oxidizer and coolant.

This simulation was no easy task. The model required 40 integrators, several hundred summers and multipliers, 38 dividers, and 32 arbitrary function generators. Three of those generators were functions of two variables.

The evaluation certified that the accelerator hardware can run applications "as is" in real time with four processors. In a comparative test, a supercomputer took over 250 microseconds, which was more than eight times slower than real time.

A Halifax simulation system equipped with the EAI Simulation Associates accelerator has been designed to perform the functions of the analog processor in a hybrid simulation. By using the accelerator, a pair of 17-foot racks costing over $1.5 million can be replaced with a single cabinet at about a quarter of the cost. An entire simulation can be programmed in floating point in a single source language.

Small in size, powerful in computer punch. EAI Simulation Associates developed a new computer simulation processor, yielding better-than-supercomputer performance.

STRESSES AND STRAINS

More than a step saver, computerized looks at stress, vibration, and acoustical properties of structures can be seen—before prototyping even begins.

NASA software engineers have created thousands of computer programs over the decades. These computer tools can design, test, and analyze a broad assortment of aerospace parts and structures. Considered one of the most successful and widely-used NASA software programs is the NASA Structural Analysis Program, called for short, NASTRAN®.

Originally created by Goddard Space Flight Center for spacecraft design, NASTRAN has been employed in a host of non-aerospace applications. NASTRAN is available to industry through NASA's Computer Software Management and Information Center (COSMIC), located at the University of Georgia.

COSMIC maintains a library of computer programs from NASA and other government agencies and offers them for sale at a fraction of the cost of developing a new program.

Computerized Structural Analysis and Research (CSAR) Corporation, located in Agoura Hills, California, is a leading producer of mechanical computer-aided engineering software. CSAR first released in 1984, an enhanced version of the COSMIC NASTRAN. This finite element analysis (FEA) program is now in use by companies around the world for solving the largest, most difficult engineering problems.

NASTRAN software mathematically analyzes a design and predicts how it will stand up under various stresses and strains. Quick and inexpensive, it minimizes trial and error in the design process and makes possible better, lighter, and safer structures while significantly reducing development time.

Established in 1982, CSAR's goal was providing the best FEA tools available. Products were successfully marketed as CSAR gained a reputation for its unique software products, training and technical support for the engineering community. In 1988, CSAR purchased the COSMIC NASTRAN source code outright, releasing the first "independent"

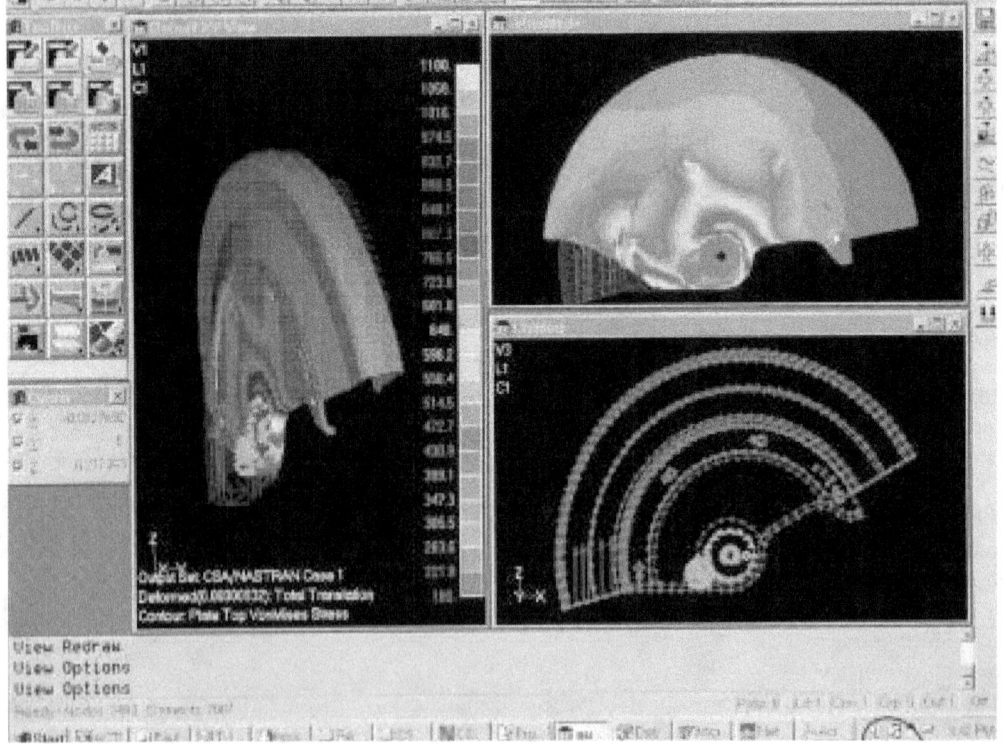

Computer software analyzes how a hardware design will stand up under temperature ranges, vibration shocks, and other stressful forces.

version of the software. All titles, ownership, and proprietary rights to the CSAR's NASTRAN version reside solely with the CSAR Corporation.

"We have concentrated on developing the capabilities that NASTRAN users find most valuable while leap-frogging our competitors with superior speed and cost effectiveness," says R. Swami Narayanaswami, CSAR's President and Chief Executive Officer. For example, CSAR has established software for general purpose nonlinear finite element analysis that can tackle everything from simple large displacement problems to complex phenomena involving impact, structural collapse, and even explosion.

A CSAR just-released software product gives engineers the ability to quickly and easily model the physical characteristics of almost any three-dimensional object, simulate static or dynamic loadings on the object, and calculate the response of the object to those loadings. State-of-the-art numerical algorithms take advantage of the computational abilities of today's high-speed workstations and supercomputers.

In 1997, CSAR celebrated 15 years in business. CSAR literally operated out of a car garage, Narayanaswami notes. He fondly remembers those early years from his office of today, in the 7,000 square foot CSAR complex.

CSAR operates a subsidiary office in London and has ongoing agreements with distributors in Germany, Japan, India, Singapore, Malaysia, and Korea. CSAR enjoys revenues from over 500 leases of NASTRAN-enhanced versions of software for the personal computer on a worldwide basis.

® NASTRAN is a registered trademark of NASA.

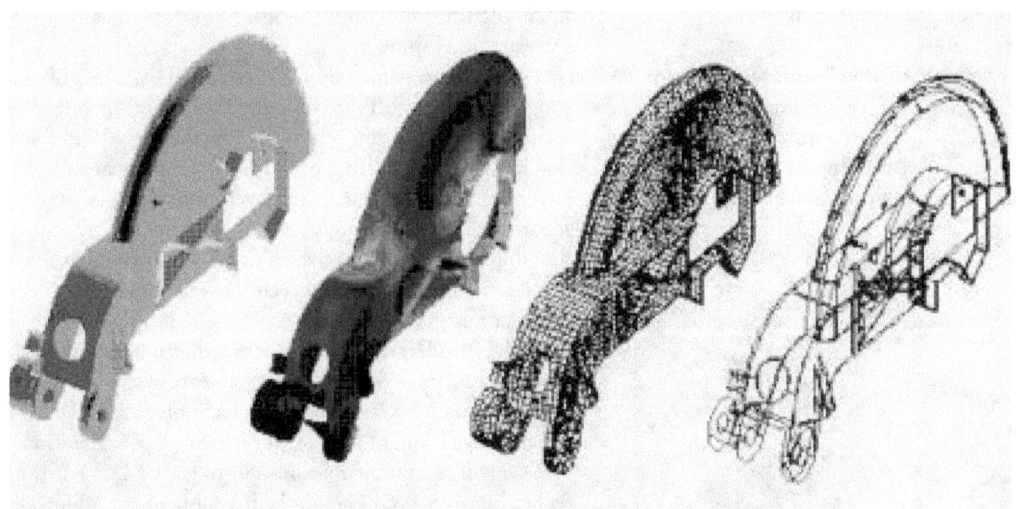

CSAR software has gained wide popularity in the engineering community. Software designed by the commercial company to perform structural tests of a component before actually being built evolved from NASA-developed NASTRAN software.

UNDER ATTACK

Another computer system break-in. Statistics show more will come. Almost 70 percent of commercial DPE/MIS organizations reported some form of information security incident in 1997. Computer data theft can result in huge money losses as computer crackers gain unauthorized access to a government agency or company's business through the Internet.

In the past, attackers have been mainly hobbyists with too much time on their hands. They are more satisfied by just taking on the sheer challenge of a computer system break-in. But attacks today have become malicious, intent on damage and piracy of intellectual property. A warning from the U.S. Department of Justice underscores the fact that the number of cyberspace criminals on the Internet will exceed five million by the year 2000.

Information security experts at Diversified High Technologies, Inc. (DHT) of Houston, Texas support the Johnson Space Center (JSC), providing information systems security safeguards.

DHT's mission statement is blunt: "Help organizations enhance productivity and protect their facilities and assets by business technologies integration without impacting ongoing operations."

In response to a 1992 NASA-wide cost saving directive, JSC's Engineering Directorate established the Avionics Software Development Environment Pathfinder (ASDEP) program. This activity evaluated Internet security technologies for mission critical facilities.

DHT experts in security systems are ready to help foil cyberspace break-in, an ever-growing security threat to businesses that rely on the Internet.

DHT has supported JSC's ASDEP program throughout its phases. For the first two years, DHT operated as the technology facilitator, providing technical coordination between vendors, contractors, and NASA. As technology facilitator, DHT helped coordinate the requirements and implementation of then state-of-the-art firewall functions such as dynamic network address translation (NAT) and encrypted private virtual networks (PVNs). DHT has recently provided support in secure technology migration path planning and the incorporation of new security technologies in response to JSC's evolving program needs.

JSC's ASDEP is now operating six "firewall" systems nationwide. A firewall is a set of components placed between two networks that collectively have these elements: all traffic from inside to outside and outside to inside must pass through the firewall; only authorized traffic, as defined by a business or government agency, will be allowed to pass; and the firewall itself is highly resistant to penetration.

The six firewall systems securely interconnect NASA and contractor facilities, using the Internet, to conduct mission critical functions for Space Shuttle onboard software development.

Knowledge gained by DHT in the NASA ASDEP program has allowed the company to successfully offer security services to the commercial marketplace. DHT has developed and is offering to both the commercial and government marketplaces, advanced information systems security services. These systems include: single sign-on, distributive computing environments, network intrusion detection, and independent security verification and certification.

The firm has developed a comprehensive information systems security program. Staying abreast of the leading edge information systems security technologies, new concepts are brought forward for customer consideration.

One of many valuable aids put to use by DHT is the technology test bed. The test bed, a subnet with a firewall system, is used to evaluate network security and connectivity technologies. That evaluation helps determine what major products are required and the corresponding integration needs. This test bed concept was directly derived from ASDEP.

DHT's goal in working with its customers is simple and direct: "To minimize the customer's Internet information security risks by providing a cost-effective, efficient, and well-implemented information systems security solution."

PLANNING FOR "WHAT IF" SCENARIOS

You think *you* have a busy schedule? Try running the Hubble Space Telescope!

Now the most successful scheduling technology ever developed for NASA is available to customers who want to minimize production costs while maximizing throughput.

The founders of Interval Logic Corporation of Sunnyvale, California were the developers of the SPIKE scheduling system specially created for NASA's orbiting observatory, the Hubble Space Telescope. The operational efficiency of the astronomical eye-in-the-sky was doubled by the scheduling software. This same system has now logged over 60,000 hours of operation in mission-critical applications, managing assets worth billions of dollars.

Interval Logic Corporation was founded in 1995 with the express purpose of providing scheduling and optimization solutions. The group had help in opening its doors, courtesy of the Ames Technology Commercialization Center (ATCC). This hightech incubator is available for companies commercializing NASA technology.

Formerly an ATCC company, Interval Logic has moved on from incubation status to a company producing powerful and sophisticated scheduling software based upon NASA technology. Software offered by the company allows scheduling and optimization of large, complex systems.

Interval Logic has introduced its SPIKE™ scheduling software on a commercial basis, targeted at the semiconductor industry. The product can dramatically streamline semiconductor production. Semiconductor manufacturing operations comprise some of the most difficult and complex of all manufacturing scheduling problems in business today. The frequent challenges of innovation in this industry require adaptation to constant change. Global competitiveness creates intense pressure on production managers to effectively manage manufacturing cycles and cut costs.

To answer semiconductor industry needs for dramatic increases in handling demand, along with optimizing fabrication and backend operations, Interval Logic has created a scheduling and optimization product family.

"Semiconductor manufacturing is not run like a newspaper or steelmill," says Donald Rosenthal, President and founder of Interval Logic. "Simply focusing on dispatch rules for a few pieces of equipment can't maximize throughput across a plant, even in the best of circumstances," he says.

Rosenthal explains that as the product mix varies, as hot orders arrive, or whenever plans are changing, adjustments to flow are needed. That assures that the entire fabrication, assembly, and test floor is optimized for throughput—while maintaining balance across the entire facility. Interval Logic's software is capable of providing a user look-see at "what if" scenarios, and does this with unmatched speed, he points out.

With a click of the mouse, lot rescheduling, machine availability, and product cycle times are effortlessly manipulated. Easy to use, the software planning and scheduling system can provide visibility and control over semiconductor manufacturing operations—from wafer release through final test.

Despite billion-dollar-plus fabrication investments, overall utilization can be as low as 30 percent. That is a tremendous cost drain, Rosenthal says. To optimize profits in these high cost fabrications, chip makers must elevate their utilization factors. "It's a pervasive challenge," he adds.

Taking a look at the business end of semiconductor manufacturing, by better production optimizing, obvious objectives are achieved—increased profitability, reduced overhead, and ensured customer satisfaction.

Interval Logic has evolved a set of products that provide a leverage—the power to act effectively—software for making hard choices easier.

™SPIKE is a trademark of Interval Logic Corporation.

Interval Logic Corporation's scheduling system is based on software created to double the operational efficiency of the Hubble Space Telescope.

ANTI-HASSLE CHIP

It is estimated that over 25 million people worldwide have Internet connections. That number is on a fast-moving growth curve daily. However, four out of five people in the United States still remain "Internet no-shows." They remain unconnected due to two primary obstacles: the lack of a personal computer and the perceived complexity of getting on the Internet.

With assistance from Ames Research Center, the iTv Corporation (iTvc), a San Mateo, California-based startup, has introduced a computer chip that minimizes the hassle of logging onto the Internet. The work by iTvc was supported by the Ames Technology Commercialization Center, an organization that "incubates" small businesses. With $900,000 in venture capital funding, raised while in the NASA Ames business incubator, the company was able to proceed with development and rollout of its first innovative product.

Although iTvc is responsible for their own funding, the collaboration with NASA gave iTvc access to the very latest NASA microprocessor technology. NASA contracted with iTvc to adapt a unique microprocessor based on an asynchronous architecture to operations in a space environment. The technology was invented by Charles Moore, one of the three founders of iTvc. Moore was also the inventor of software language used extensively on NASA space probes and satellites.

The company's first product to graduate from the incubator was built to link the Internet to the television, providing a high quality image for electronic mail (email) and World Wide Web browsing. The device uses a proprietary 400 MIPS (millions of instructions per second) throughput microprocessor chip, which was developed during iTvc/NASA collaborative projects.

Instead of using a personal computer, iTvc hardware uses an existing television and telephone line to connect the user to the Internet. With 98 percent of all U.S. households owning a television, the potential market for the equipment is large.

The Internet access hardware provides true plug-and-play. Nothing to configure; no unintelligible computer commands. A connecting of cables, a push of the button, and the connection is made. The system is powered by a breakthrough microprocessor, originally developed for NASA space missions, then adapted by iTvc for fast, powerful yet inexpensive Internet access. A 101-key keyboard is included as standard equipment with the iTvc system.

The company has developed a full custom microprocessor that will enable access to the Internet through an inexpensive device and keyboard. The 100-MHz chip uses a 20-bit wide internal data bus to process 4 instructions per clock cycle. The power of the processor is low, some 90 milliwatts. That is a fraction of the memory and power of conventional microprocessors. Although the primary application of the microprocessor is for Internet access applications, other data path options and coprocessor arrangements can be implemented.

The microprocessor is commercially in quantity production, supported with a compliment of design tools for customization and adaptation as either a licensable core or as a complete microprocessor. In addition to the microprocessor, a full multi-tasking operating system and applications software are available for complete Internet access—World Wide Web browsing and electronic mail.

Other uses of the microprocessor include cell phones, DVD players, cable modems, video conferencing equipment, digital cameras, and wireless local area networks and wide area networks. In general terms, the iTvc microprocessor supports at least a $100 reduction in costs of mass market consumer electronics, relates Gary Langford, President of iTvc. That means big business, he says.

"A forecast for the Internet is its use by almost every household in the United States within 15 years, with half of all homes having the Internet in 6 years," Langford says. The reality of this forecast will only happen if the Internet devices are inexpensive and easy to operate, he adds.

In 1997, iTvc raised several million dollars in venture capital. First trial demonstrations of the company's Internet appliance started in full swing. The company is currently working in partnership with Korean and Japanese electronic manufacturers to design new, low-cost consumer products. NASA has expressed interest in iTvc's computer prowess for a set of Earth observing satellites and the search for life on Mars, to handle massive streams of data broadcast from these spacecraft missions.

Smarts on a chip. This multiprocessor is the brain behind iTvc's low-cost approach to making the Internet available through the television. The micro-processor is adapted from electronics technology designed for robotic Mars missions.

ENGINEERING PERFORMANCE

An essential tool for designing distributed computer systems has been developed by AST Engineering Services, Inc. (AES) of Englewood, Colorado.

Under Small Business Innovation Research (SBIR) contracts with both Goddard Space Flight Center (GSFC) and the Jet Propulsion Laboratory (JPL), AES conceived of a system engineering computer software tool called QASE® that can model how certain applications will affect a proposed system's performance. QASE stands for Quantitative System Engineering, and is a commercial product based on the merger of two ancestor tools from the JPL and the Navy.

The product is currently used to support NASA's Earth Observing System (EOS) Data and Operations System (EDOS) initiative. EOS satellites are part of a 10-year effort to discern human impact on global climate change. Such research is destined to underscore the need for better stewardship of the Earth's delicate biosphere. Computer power necessary to assemble and then distribute massive databases of global climate change information is mind-boggling.

In a world where governments and large corporations rely on massive computer systems to operate, one mistake in developing a system can cost millions, if not billions of dollars. Enter AES, five years of work, and the development of a systems-performance tool. At the root of the software tool is a single tenet: engineering more performance into software instead of throwing hardware at the problem.

AES' QASE computer software tool allows its user to describe and analyze a complex computer system, and evaluate system timing and capacity. Easily operated by the user, being a modeling and simulation expert is not required. The system description is automatically translated into analytics, simulation models, and executed.

This impressive software tool is a hierarchical entity-attribute specification. Object types provide for efficient and consistent descriptions of hardware, software, and data. Entities include:

- *Hardware Diagrams* - graphical processor, storage, and communication architectures;
- *Software* - structures definition of resource use;
- *Data* - logical data stores and flows;
- *Operating Systems* - process overhead and scheduling disciplines;
- *Communication Protocols*; and *Allocations* - software to processors and data to storage devices.

AES promotes its NASA SBIR-funded product as giving users key benefits. Among them are to:

- Respond with confidence in proposals and reviews that your system design meets performance requirements;
- Reduce costs for the users and customers by knowledgeably sizing and using computer network resources;
- Compare system performance among multiple vendors and different architectures;
- Streamline the software design process by identifying those components that most affect performance; and
- Redesign and troubleshoot current systems.

As the era of the International Space Station begins in 1998, AES has already provided system engineering expertise to NASA, in a subcontractor role to IBM Federal System Division. The company supported the development of a Space Station Data Management System (DMS) through analysis and requirements definition. The Space Station's DMS was reviewed as to requirements for traceability, completeness, and testability. A second task involved examining the overall Space Station objectives and operations. AES work in this area resulted in identifying which Space Station activities could be automated on the DMS, and defining the technologies for doing it. The study resulted in a report detailing 20 automation candidates across the spectrum of Space Station operations.

®QASE is a registered trademark of AST Engineering Services, Inc..

ASE computer software tool models how certain applications will affect a proposed system's performance. NASA SBIR funds brought the software system to fruition, now commercialized and applied to major NASA endeavors, like the Earth Observing System and the International Space Station.

IT'S A CLEAN MACHINE

Cleaner running, environmentally friendly auto engines and gas turbines, as well as improved fuel efficiency—that is a nice sounding combination.

Since its founding in 1986, Precision Combustion, Inc. (PCI), New Haven, Connecticut, has focused on development and commercialization of its proprietary, clean, and efficient combustion and air pollution control technologies. The company reports it has invested in excess of $10 million in research and development contracts obtained from government and industrial customers in the development of its products.

Small Business Innovation Research (SBIR) monies from Lewis Research Center were awarded in 1986 and 1990 to PCI. The research proved the viability of efficient, cost-effective catalytic reduction of gas turbine nitrogen oxide emissions along with fuel efficiency.

PCI's commercialization efforts are focused on two major product lines, a catalytic combustor for gas turbines and a catalytic converter for automotive applications.

The use of catalysts inside the combustion chamber allows leaner combustion which reduces the formation of nitrogen oxides while preserving the efficiency of advanced combustion turbine designs. Gas turbine catalytic combustion technology offers emission reductions and cost savings compared to more established low emission technologies such as lean premixed combustion and selective catalytic reduction in meeting gas turbine emissions regulations.

Early Lewis Research Center interest in PCI's abilities has paid off handsomely. The company has developed an Advanced Technology Catalytic Combustor (ATCC), and the Microlith, an automotive catalytic converter. "These technologies would not have been commercially feasible without the support of the NASA SBIR program," says PCI president, Kevin Burns. "The follow-on commercial support that was built directly on this early research has enabled the company to grow substantially," Burns adds.

"Our catalysts inside the combustion chamber allow leaner combustion, avoiding formation of nitrogen oxides while preserving the efficiency of advanced combustion turbine designs," observes PCI chief scientist, William Pfefferle, the inventor of the original catalytic combustor for gas turbines. "We are integrating advanced catalytic technology with new combustor designs. Ground power generation customers are going to like this technology because it combines clean emissions with efficient high firing temperatures along with stability, operability, and reliability," Pfefferle concludes.

The combustor technology has potential application for aircraft turbine engines, both for military and commercial markets. In November 1996, a long-term business agreement to develop, manufacture, and sell new catalytic combustor products for Westinghouse Power Generation machines was announced. In a step toward broader industrial commercialization of catalytic combustor technology, PCI was awarded a $750,000 contract from the U.S. Department of Energy. The work is geared to enable gas turbine manufacturers to meet new and stricter environmental regulations.

The development of the Microlith automotive catalytic converter (left) for ultra-low exhaust application turbine engines was advanced by a Lewis Research Center Small Business Innovation converter Research contract. Also developed were the stand-alone converter (right) and cartridge (foreground).

NASA's SBIR program has also supported another important catalytic product under development at PCI. In 1994, Marshall Space Flight Center began an SBIR program developing PCI's ultra-compact catalytic converter for spacecraft life support. The company has been concurrently developing this technology as a compact, lightweight, and high efficiency catalytic converter for automobiles. Prototypes are in test with major U.S. auto manufacturers. Recent testing of the converter has demonstrated emissions reduction that exceeds the new ultra-low emission vehicle standards of the day. The Environmental Protection Agency (EPA) is providing PCI ongoing commercialization support for its converter. PCI has been selected as one of only five small businesses included in an EPA effort, the Environmental Technologies Initiative Program. PCI is currently exploring strategic alliances for scaling up production and developing a pilot manufacturing plant.

PCI anticipates its current growth and financing efforts will enable it to become a highly value-added, fully-integrated manufacturing company, providing core catalytic components and products for a range of clean combustion market opportunities. Other products in the development pipeline include a durable catalytic grow plug for diesel and gas turbine engines, and hybrid electric vehicle and fuel cell components, which provide a long-term stream of proprietary technology products fueling future company growth.

Dr. Paul Menacherry, a PCI research and development engineer, is working with a custom-built test rig used to simulate automotive exhaust gas.

LOW LEVEL LEAKS

Getting "up close and personal" with a leak at a distance demands advanced technology detection equipment. A portable long-range leak detection module was designed at the Kennedy Space Center (KSC) to spot jet-type leaks in the fluid systems of critical, out-of-reach launch and ground support equipment.

The long-range leak detector improvements were born when the entire Space Shuttle fleet was grounded due to leak problems in the main engine compartments. KSC contract engineers devised off-the-shelf ultrasonic leak detectors. While these detectors proved helpful in finding the leaks, they lacked long-range sensitivity, simple operator interface, and ease of use. Further enhancement of the detectors was called for when a set of Space Shuttle solid rocket boosters demanded a leak check once-over.

The KSC-developed leak detector has been refined for commercial purchase. This innovation is being offered under an exclusive licensing agreement between NASA and UE Systems, Inc., Elmsford, New York.

Working like an ultrasonic telescope, the Ultraprobe 2000 hand-held unit incorporates the long-range, NASA-developed module. It incorporates state-of-the-art circuitry, improved transducers, and a unique collecting horn that delivers a higher degree of reliability, sensitivity, and versatility over previously used leak detectors. UE Systems offers the Long Range Leak Detection Module (UE-LRM-2) and the Close Focus Leak Module (UE-CFM-2).

A uniquely shaped receiving chamber resembles the intake port of a jet engine in the UE-CFM-2. The shape enhances the turbulent flow of a pressure or vacuum leak and gives a user the capacity to locate minute leaks that might otherwise go undetected. Some of the most common plant applications are in leak detection in pressure and vacuum systems, such as boiler, heat exchanger, condensers, chillers, distillation columns, vacuum furnaces, and specialty gas systems. The most subtle changes in the integrity of seals and gaskets in tanks and pipe systems can be exposed.

"The new system," says Terrence O'Hanlon, president of UE Systems, "was developed to overcome the limitations of low sensitivity and discrimination against background noises found in most other commercially available instruments."

In the UE Systems' Long Range Leak Detection Module, the unique parabolic design of the sonic collection chamber is shaped to reflect all ultrasonic signals directly to the apex of the instrument's transducer. The signal is then preamplified and transferred to a pistol housing where it is amplified further, then mixed with a signal from a local oscillator that shifts the received ultrasonic signal to an audible frequency. In electronic terms, this process is commonly referred to as hetrodyning or demodulation. The captured, "double amplified" audio signal is then sent to headphones or a spectrum analyzer for further analysis or data storage.

O'Hanlon explains that the UE Systems leak detection device could save American manufacturers money. He points to a Department of Energy study that determined in most factories, there is an opportunity for energy conservation. In some factories, the study found an "outrageous" waste of energy associated with compressed air systems. The study states that detection of one air leak could save a firm thousands of dollars each year. "Leaks equal money," O'Hanlon asserts.

The unit is small, lightweight, and rugged compared to other currently used systems, opening the door to a wide variety of commercial uses. A multifaceted instrument, the detector provides testing capabilities ranging from simple leak detection to sophisticated methods of mechanical analysis. The ultrasonic device is an ideal center piece of predictive and preventive maintenance programs in virtually every industry. Commercial application of the detector hardware includes: pipelines, underground utilities, air-conditioning systems, petrochemical systems, power transmission lines, and in medical devices.

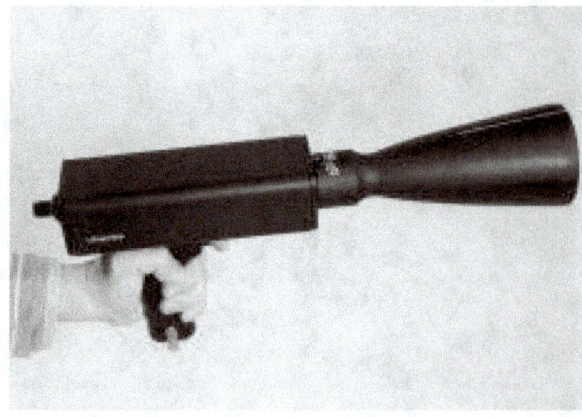

The most subtle of leaks can be discerned with UE Systems' is detection equipment. This hand-held device was born out of NASA's need to detect leaks in out-of-reach Space Shuttle hardware and launch pad equipment.

FROZEN SMOKE

Aerogel is the lightest solid material known. Some call it "frozen smoke"...others see it as "pet cloud."

Despite its density—three times that of air—the material has tremendous insulating capability. NASA scientists have found aerogel critical for several space missions. Part of the Mars Pathfinder mission that landed on the Red Planet in July 1997 included treks of the tiny Sojourner rover. To guard against the rover freezing in the chill of a Martian night, aerogel kept Sojourner on the run. Aerogel is also set to fly on the Stardust mission, being readied for launch in 1999. Stardust is built to haul aerogel that will snag dust samples from a comet and return the specimens to Earth.

Under a NASA Small Business Innovation Research (SBIR) contract with the Kennedy Space Center (KSC), Aspen Systems, Inc., Marlborough, Massachusetts, has manufactured a variety of aerogel products, including aerogel superinsulation. Its advantages are more than clear. A one-inch thick aerogel window has the same insulation value as fifteen panes of glass and trapped air.

Aspen Systems' aerogel-based superinsulation is an innovative, flexible cryogenic insulation with extremely low thermal conductivity. The design of this product takes advantage of the unique properties of specially-made aerogel materials. Aerogels formed at the fiber-fiber contacts of a matrix material force solid heat transfer to occur through the very low thermal conductivity aerogels. Air conduction is greatly reduced due to the very fine pore size of the aerogels.

Flexible aerogel insulation was produced for NASA in two separate products: one for cryogenic and room-temperature applications, and the other for high-temperature uses. The flexible cryogenic superinsulation was developed for KSC, while the high-temperature version was designed to Ames Research Center specifications.

Founded in 1984, Aspen Systems has matured into a multi-disciplinary organization with a number of products ready for commercialization. Company expertise ranges from energy and environmental systems to specialty materials, photonics, and biotechnology. As a technologically diversified small business, Aspen Systems serves many Federal and state agencies, as well as industrial and commercial clients.

First steps in commercializing the easy-to-use aerogel superinsulation have taken place, notes Hamed Borhanian, Vice President of Aspen Systems. "We are now in the process of securing investor funds to embark on a full scale commercialization," he says.

Superinsulation can be manufactured as fully flexible or as relatively rigid, but not brittle, for insulating a wide variety of objects. An external jacket of pipe, foil, or plastic may be employed, as required, to protect superinsulation from environmental or mechanical impact. SBIR awards to Aspen Systems has spawned a unique aerogel fabrication process, in which the physical properties and thermal performance of the aerogels can be tailored for a given application.

Since the moment Aspen Systems opened its web site advertising aerogel material, hundreds of inquires have poured into the company. This interest has shown the vast potential of aerogels. Potential markets appear ripe for commercial exploitation. Aside from the cryogenic insulation market that constitutes the first tier market for aerogel insulation, other product areas also look favorable. Among these: insulation for offshore oil well underwater pipelines; insulation for shipping containers; refractory insulation for automotive firewalls, floorboards, and exhaust systems; as a high-efficiency filtering media; and as acoustic damping insulation for buildings, process equipment, and headphones.

This smoky hue of a product appears to have a clear future.

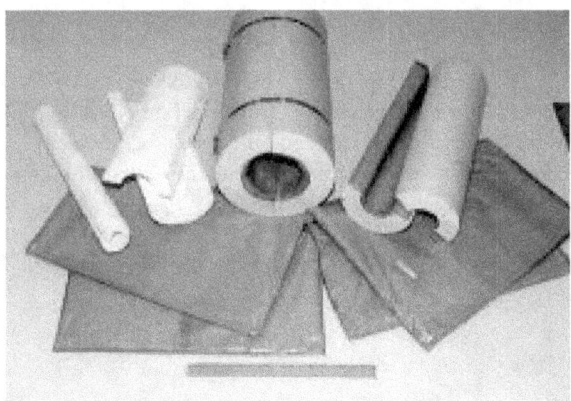

Blankets of aerogel materials act as phenomenal insulators. NASA-supported research has stirred commercial interest in the product for a variety of business and consumer applications.

THIN FILM...LARGE PAYOFF

Good things come in small packages. In the case for space, ultra-thin film can be tightly packaged for launch, then deployed in orbit to establish large solar concentrators and antennas for geographical surveys and communications. Lightweight, inflatable structures will soon act as huge reflectors that focus the sun's energy to heat propellant for thrusting payloads to high altitudes. Gossamer-like thin film might take the shape of collecting and transmitting dishes for 21st century satellites that collect solar energy, then beam the energy to Earth.

SRS Technologies, Huntsville, Alabama, has been working on the cutting-edge of thin-film concentrators for over ten years. A progressive series of Small Business Innovation Research (SBIR) contracts awarded to SRS has come from Langley Research Center and Marshall Space Flight Center. In particular, Langley Research Center funded SBIR awards focused on the use of large thin-film concentrators for large space-based antennas. SRS has licensed and now produces commercially several NASA-developed polyimides—a colorless, low dielectric, radiation-resistant, and moisture-resistant material developed for high-temperature applications. "Through the SBIR programs, SRS Technologies has been able to develop a unique competitive advantage and provide solutions to difficult problems," says Harold Pastrick, Corporate Vice President of SRS Technologies.

These polyimides' transparencies are superior to competing polyimides. The materials solubilities provide design and processing flexibility. NASA developed the materials to meet requirements for a transparent colorless

material that exhibits extreme thermal stability and greater resistance to radiation and atomic oxygen in low Earth orbits.

SRS now manufactures under license the NASA-developed polyimides as powder, resin, and rolled film. These materials, known as

SRS Technologies currently produces two polyimides, the LaRC™-CP1 and LaRC™-CP2. These polyimides offer many advantages over commercially available polyimide materials, including thermal stability, radiation resistance solubility, and transparency.

LaRC™-CP1 and LaRC™-CP2, have exhibited good long-term storage capability, which enables SRS to produce large inventories. SRS has had great success applying the materials to thin-film deployable concentrator/antenna technologies. A 16-foot on-axis antenna was built for Langley Research Center. One recent test demonstrated the thin film's characteristics to form a 23-foot diameter antenna dish. It weighed in at an impressive seven pounds. SRS has spun-cast films up to several feet in diameter, while larger films have been solution-cast.

These SRS polyimide's can be used in laminates, films, molded parts, and stock shapes. Thickness of the thin film can widely vary, ranging from 0.3 mils to 2.0 mils, up to 24 inches in width. SRS plans to increase production of film from 48-inch to 60-inch widths in the near future.

SRS's production and use of the NASA-developed polyimides have positioned SRS for promising commercial opportunities. Those promising business opportunities include production of rolled-film material using an SRS-designed machine, made possible through the SBIR program. Currently, SRS is the only company in the world licensed to produce this unique product commercially.

NASA's Technology Applications Team at Research Triangle Institute (RTI) has worked closely with the Langley Research Center to assist in providing in-depth analysis of the potential applications and market factors affecting the commercial viability of the materials. RTI has helped NASA market the technology and develop licensing and cooperative development agreements with interested companies. SRS Technologies signed an agreement with NASA, and is currently the only company licensed to produce the colorless polyimides.

Five major applications for the polyimides were identified and studied in detail by RTI. These were: flat panel displays, microelectronics, coatings, solar arrays, and thermal control materials. RTI's efforts helped to facilitate the SRS license and cooperative development efforts.

Additional applications for these temperature and radiation resistant, colorless, and/or low dielectric polyimide materials may include: flexible printed circuit substrates, high temperature wire and cable wrapping, electric motor and generator insulation, and possibly protective coatings for art and outdoor statues. Several companies are currently assessing the thin-film technology for specific applications.

™LaRC-CP1 and ™LaRC-CP2 are trademarks of SRS Technologies.

TOUGH LIKE METAL

A new coating material can greatly extend the lifetime of ceramic composites, making them more than 1,000 times more durable.

Lewis Research Center has awarded Small Business Innovation Research (SBIR) contracts to Advanced Ceramics Corporation (ACC) of Cleveland, Ohio, as part of a NASA campaign to study propulsion materials within its high-speed aircraft program.

ACC has developed a family of high temperature and specially treated boron nitride coatings. These interface coatings circumvent moisture that complicates composite manufacture and ultimately degrades composite performance at high temperatures. The moisture-resistant coatings extend the shelf life of composite materials and resist thermal oxidation in moist air. Ceramic fibers woven into fabrics can now be coated with this advancement. The coating work is seen as a boon to the ceramic composites industry.

What are ceramic composites? They are a new type of material made by reinforcing refractories with high-strength ceramic fibers. They are as strong as metal, yet can withstand higher temperatures. One immediate use for composite materials is use in turbine engines, an application that can make the engines far more efficient.

Industrial gas-fired power turbines that have composite combustor liners and shrouds can compress fuel and air at higher temperatures than all-metal turbines, which increases efficiency. The turbine can produce more kilowatt-hours from the same fuel than it could if metal combustor liners and shrouds were used. This is a big advantage in the competitive and very cost-conscious power turbine industry.

Aircraft turbine engines which have composite liners and shrouds also weigh less than all-metal turbines. This allows the aircraft to carry a larger payload or fly a longer distance using the same fuel. "More payload and more distance from the same engine and the same fuel" is the mantra of civil aviation.

Why did ceramic composites require better interface coatings in the first place? A sliding interface between fiber and ceramic allows mechanical loads to be distributed to the reinforcing fibers, making ceramics "tough" like metals. A lot of energy is needed to produce a fracture in a ceramic composite. But conventional interface coatings are removed by air, especially by moisture in the air.

Through NASA SBIR-supported research at Advanced Ceramics Corporation, a high-performance barrier coating was made, capable of being applied to composites by chemical vapor deposition. This new coating enables continuous ceramic fiber composites to be used at higher temperatures, permitting development of ceramic combustors and hot parts for advanced turbine engines. Non-aerospace applications of the barrier coating include stationary power turbine components, radiant tube burners, gas-flow heat exchangers and filters, and waste incineration systems. ACC received a patent for its work on interface coatings for ceramic fibers in January 1997.

Advanced ceramic materials are experiencing rapid growth because they perform extreme engineering and structural functions that are outside the performance range of metals and metallics. Advanced ceramics operate effectively from super-cold to super-hot temperatures. Paradoxically, these materials are excellent electrical insulators and also perform as superconductors.

ACC's investigations into special coatings for ceramic composites are expected to provide an entire new range of high-performance material options. These materials are making important contributions in such leading technologies as heat engines, integrated circuit manufacture, and superconductivity. They also are important in manufacturing and processing applications such as molten metal handling and glass making.

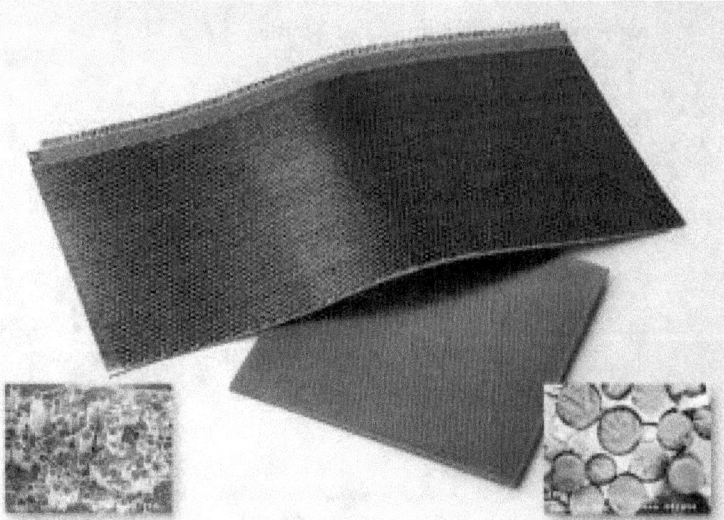

Advanced Ceramics Corporation research makes composite materials more durable by application of patented coatings. Work was stimulated by NASA interest in future high-speed civil air transportation.

SOMEWHERE IN THE RAINBOW

Spray combustion processes can be studied, thanks to a device built to measure fuel droplet size distributions by bathing them in laser light.

Aerometrics, Incorporated of Sunnyvale, California, first developed, through a Small Business Innovation Research (SBIR) contract with Lewis Research Center, a particle analyzer using a Phase Doppler technique. This laser instrument provided a non-disruptive method of determining particle size, number density, flow velocity, diameter versus velocity, trajectory, turbulence intensity, and more.

A follow-on NASA SBIR contract with Aerometrics has now built upon that earlier product. The new technology can be integrated with the company's Phase Doppler Particle Analyzer (PDPA) systems, or used independently.

This latest device from Aerometrics—a rainbow refractometer—is keyed to measuring the refractive index of a droplet by monitoring, in a non-intrusive manner, the main rainbow location with a Charge Coupled Device (CCD) linear array. Refraction is the bending of a ray of light at the boundary of two mediums of dissimilar nature. When droplets are illuminated by laser light, they give rise to a characteristic rainbow pattern in the backscatter direction. This pattern can be recorded using a CCD line camera and the data processed to extract the refractive index information, which is a function of droplet temperature, for a given droplet composition. The refractive index of particles as small as 20 microns in diameter can be measured. For particles less than 150 microns, the Doppler particle analyzer and the refraction instrument can be used in conjunction.

Some applications of the technology include:

- **Spray Flame Characterization**
  Fuel droplet size and velocity in complex spray flames can be measured. No other instrument is capable of measuring individual fuel droplet temperatures in a spray flame. By measuring droplet size, velocity, and temperature, data can be used to validate advanced theoretical spray combustion models that are routinely used in the design of fuel efficient combustors.

- **Droplet Combustion Studies in Microgravity**
  A compact diode laser-based rainbow refractometer system is under development by Aerometrics. This instrument can be used for obtaining the temporal evolution of fuel droplet size, internal temperature profile/gradient, and droplet regression rate in single droplet combus-

tion experiments. Those tests are to be carried out aboard NASA's microgravity producing aircraft, created by parabolic flight trajectories of the plane.

- **Rocket Engine Mixing Studies**
  In liquid rocket engine injectors using more than one liquid propellant, adequate and uniform mixing of the fuel with an oxidizer is essential if efficient combustion is to be realized. Tasking the combined Phase Doppler/rainbow refraction device provides a useful means for studying liquid-liquid mixing even in a reactive environment. Individual droplets within propellant sprays can be sorted out based on the measured droplet refractive index. This information can subsequently be used to compute the spatial and temporal variations of the liquid propellant mixture ratio.

Devices built by Aerometrics are expanding the boundaries of particle diagnostics technology. Not only research oriented work by the government, industry, and educational institutions can benefit by using the laser equipment.

"Everyone benefits from research geared toward furthering the quality of life," says marketing specialist for Aerometrics, Kristi Altier. "Our rainbow refraction instrument will make important contributions toward important chemical and combustion research," she says.

Aerometrics' instrument can inspect, via laser light, individual droplets of fuel for combustion research.

OPTICAL MICROMACHINING

Rapid fabrication of optical elements has been demonstrated using laser micromachining.

Potomac Photonics, Inc., Lanham, Maryland, is a manufacturer of ultraviolet excimer lasers and precision laser micromachining workstations. The company's forte in these areas, and its ability to develop and commercialize laser and micromachining technology has been aided by NASA Small Business Innovation Research (SBIR) awards.

Marshall Space Flight Center endowed Potomac Photonics with an SBIR award to construct and demonstrate a unique tool that fills a need in the area of prototyping diffractive and refractive micro-optics. This tool is an integrated computer-aided design and computer-aided micromachining workstation that extends the abilities of optical designers on a scale not before possible.

With a single computer-based system, an engineer is able to design, optimize, manufacture, measure, and refine optical elements of arbitrary structure on virtually any two-dimension or three-dimension surface. At the same time, the simplicity and affordability of this tool places it within the reach of many designers and fabrication specialists of optical components. Many in the optical design industry are now unable to exploit diffractive elements in their projects due to the staggering development costs associated with conventional microlithographic methods.

Advances in semiconductor processing techniques have made diffractive optical elements a viable alternative for many applications. However, standard fabrication methods are complex, expensive, and time consuming. Thus, there has been considerable effort in the development of fabrication techniques that are better suited for prototyping and quick turnaround. Some approaches to inexpensive fabrication of diffractive optical elements include the use of desktop publishing software and commercial imagesetters, direct-write laser exposure of photoresist, and production of gray level masks.

By way of NASA SBIR funds, Potomac Photonics began examining excimer laser use to directly ablate surface relief into polymer and glass materials for the production of diffractive and micro-optical components. Ultraviolet laser ablation allows for controlled material removal rates and surface finishes of optical quality. When compared to other techniques, ultraviolet laser ablation is advantageous in the reduced number of processing steps required for fabrication.

The method prompted by Potomac Photonics research makes use of a compact, high repetition rate pulsed excimer laser integrated into a workstation that is suitable for fabrication of diffractive optical elements. This system allows the user to design, fabricate, and refine the diffractive structure. Practical diffractive optical

elements can be produced in minutes to hours. That is a fraction of the time typically required by more conventional microfabrication techniques.

This advancement is likely to dovetail nicely with the commercialization of other products at Potomac Photonics. Commercialization efforts have been rewarded by strong growth since its founding in 1982. Employee growth is an indicator of the company's business in excimer lasers and micromachining technology. Today, the bulk of its business is the sale of turnkey laser micromachining systems, with units now all over the world. These workstations use laser ablation to fashion micron-scale structures directly into the substrate material. The applications span a wide array of functions, from dicing and hole-drilling in flex circuits to customizing high-technology medical devices.

Potomac Photonics considers the rapid fabrication technology to have a number of immediate applications, including sensors and monitoring equipment, analytical instruments, fiber optic distribution and communication, neural networks and optical computing, lasers and laser instrumentation, pattern recognition, displays, and information storage.

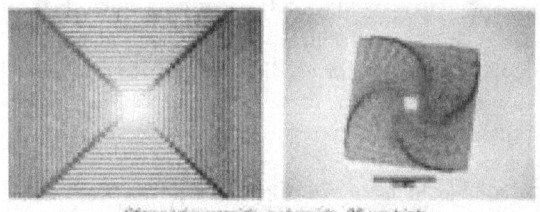

Stepped pyramids, polyimide, 25 µm high

Micromachining of optical components can be an effortless task using excimer laser technology. A new system under development cuts production costs and offers faster fabrication times over conventional micromachining techniques.

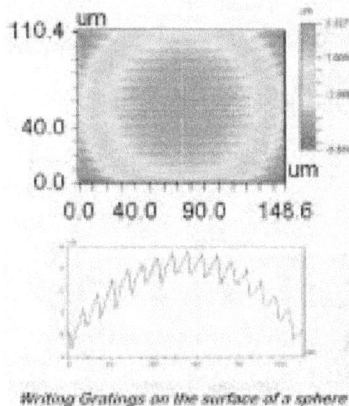

Writing Gratings on the surface of a sphere

Rapid fabrication of optical elements can be accomplished using laser ablation controlled from a workstation.

PROTOTYPING THE FUTURE

Advanced Ceramics Research (ACR) of Tucson, Arizona, has been labeled "The Idea Factory"—and for good reason.

This young up-start company, just a handful of years old, is already the 11th fastest growing high-tech firm in Arizona. ACR has taken on the mantle of research in advanced materials and process development, transforming scientific concepts into technological achievement. Doubling in size, its equipment-packed, state-of-the-art facilities demand that ACR ready itself for yet another expansion.

Through a Small Business Innovation Research (SBIR) award from Marshall Space Flight Center (MSFC), ACR developed a high pressure and temperature fused deposition system. This ACR-designed rapid prototyping system is known as extrusion freeform fabrication. Using solid feedstock, this system can extrude and form shaped components, using solid powders and/or chopped fibers as fillers to the thermoplastic feedstock. One near-term application for the equipment is in the biomedical prosthesis business.

Three-dimensional rapid prototyping is a process in which physical models are quickly and inexpensively created directly from computer-generated models. Rapid prototyping decreases the product development cycle necessary to compete in the market places of today. Concepts can be transformed into inexpensive prototype parts within days instead of weeks; designs can be verified and engineering costs decreased—no surprises at production time, despite difficult-to-do designs.

Durable, high-strength thermoplastic used for rapid prototyping gives models increased functionality. It provides impact resistance, toughness, heat stability, chemical resistance, and rigidity to permit functional tests on sample parts. Its properties also allow for post-processing techniques such as machining, drilling, tapping, painting, glueing, and sanding.

In another area, Marshall Space Flight Center also contracted ACR to fabricate a set of ceramic engines, to be appraised for a solar thermal rocket engine test program. Along

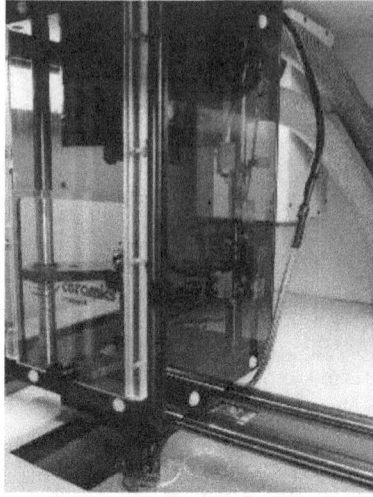

High pressure retrofit modeler transforms computer-based digital images into prototype hardware to ensure customers of a quality final product.

with the MSFC work, ACR received SBIR funding through Ames Research Center. That work centers on continuous fiber reinforced composites for heat shield applications. The technical success of this project, and spinoff work in ceramic plate manufacturing, have attracted several commercial clients and other federal customers, ACR reports.

ACR has developed and implemented three key strategies for maximizing the commercialization of its SBIR programs. These are: Case 1 - the exploitation of the directly-developed technologies from the SBIR program; Case 2 - taking advantage of the credibility associated with both the award and work performed for a Federal SBIR program; and Case 3 - utilizing spinoff technologies from the SBIR work as the program is developing such technologies.

This last point is well taken. Research and development funding from NASA and other SBIR sources equaled, at one point, a total of $750,000. Those dollars have helped refine ACR's true bread-and-butter product, commercial carriers. Sales now total over $10 million. ACR manufactures world-class quality carriers for the grinding, lapping, polishing, and wafer polishing industries. The firm's commercial carrier production represents nearly fifty percent of the worldwide consumption. In other terms, one out of every two new personal computers uses hard drive disks made by ACR carriers. SBIR contracts from Marshall Space Flight Center were instrumental in initiating ACR's commercial business. Since its start, ACR has evolved from its customer base of one hundred percent Federal work to twenty-five percent Federal and seventy-five percent commercial.

Would you expect any less from an idea factory?

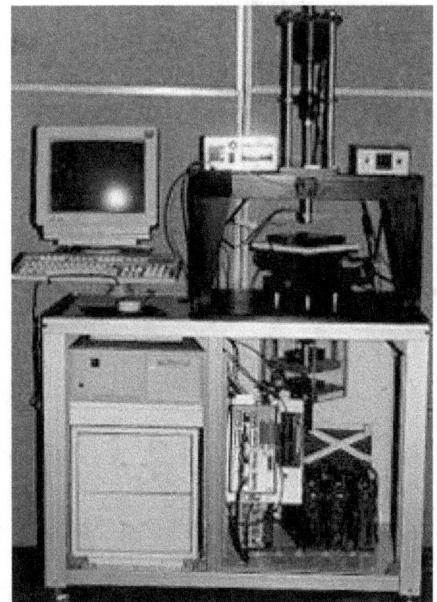

ACR equipment can shorten a customer's product development cycle and save on engineering and production costs—from composite brake pads for bikes to satellites.

A MEASURABLE DIFFERENCE—BRIDGE VERSUS LOOP

An electronic measuring circuit delivers accuracy far beyond previous methods, preventing errors caused by variations in the wires that connect sensors to data collection equipment.

This signal measurement and conditioning technology was developed at the Dryden Flight Research Center. The man behind this circuitry innovation, also called "the Anderson loop," is Karl Anderson, a former NASA Dryden measurement systems engineer. Anderson took head-on, the problem of eliminating measurement errors that arise when lead wires pass through severe environments, especially large temperature variations.

The electronic measurement circuit was patented by NASA in 1994. The Anderson loop is in use at many NASA centers where traditional strain gauge and resistance temperature detector devices are installed. It has the potential for use in many medical and industrial applications. Such devices as accelerometers, load cells, and transducers measuring torque, pressure, and temperature, even in harsh conditions, can benefit from the invention.

As for the Anderson loop, the invention proved far better than the classic, vintage 1840s Wheatstone bridge circuit design. That circuit has a long history of successfully being used to measure electrical resistance and small changes in that resistance. But it had its drawbacks. In essence, the Wheatstone bridge is a passive subtraction circuit. Anderson's approach uses a series circuit with continuous active subtraction typically accomplished in the signal-conditioning circuitry at a distance from the sensor.

Dryden demonstrated the Anderson loop, flying special "conditioning cards" as an experiment on the F-16/XL2 aircraft's Surface Laminar Flow Control (SLFC) project. Strain gauges were mounted on the plane and comparative data taken. The difference in using the Anderson loop and Wheatstone bridge methods was, indeed, quite significant. The Anderson loop provided double the signal output.

Trig-Tek, Inc., Anaheim, California, has packed Anderson loop technology into a commercial, eight-channel unit. The company is a leading builder of test equipment and data acquisition instruments. Each individual channel of the portable product using the Anderson loop offers adjustable gain, balance, and current,

as well as gauge, reference, and output test points. The equipment offers twice the fundamental sensitivity achieved in a Wheatstone bridge circuit and can tolerate random changes in the resistance of any and all lead wires.

What is behind the Anderson loop? It consists of two or more impedance elements in a series "loop" circuit. Either alternating current or direct current excitation may be used. Subtractors, the unique feature of this technology, compare the voltage drop across each sensor element with other sensor elements and/or a reference element.

A Wheatstone bridge provides a single, often nonlinear electrical function of a set of up to four sensing elements. In contrast, the Anderson loop can provide independent linear outputs from each sensing element, as well as the sums and differences of their changes. This innovation opens entirely new possibilities for "smart" transducer designs.

An attractive feature of this technology is that existing transducer designs can be converted to use Anderson loop signal conditioning for improved performance.

In 1995, the first non-exclusive license to market all or parts of the invention was issued to Valid Measurements™, founded by Anderson. The license issued allows the retired NASA employee to sell either parts, subassemblies or complete units to potential users, including manufacturers who could then sell products containing the Anderson loop on the wholesale or retail markets without obtaining a separate license from NASA. Licenses can also be obtained from NASA by other individuals or businesses to use the Anderson loop.

™Valid Measurements is a trademark of Valid Measurements.

Trig-Tek measurement device incorporates the NASA-invented Anderson loop electrical circuit. This measuring circuit is far more accurate than commonly used test procedures.

HIGH TEMPERATURE SEMICONDUCTOR PROCESS

You may not be aware of their presence, but you are likely benefiting from them on a daily basis. Thin film processes are pervasive throughout government, industry, and the consumer world; used in computer read/write heads, semiconductor, and flat panel display manufacturing.

A sputter deposition system for depositing thin films was developed by CVC, Inc. of Rochester, New York, with the support of Jet Propulsion Laboratory's (JPL) Small Business Innovation Research (SBIR) program. The SC-4000 deposition system was developed for JPL to produce high quality films of high temperature superconducting material for microwave communication system components. The system was also used to deposit ferroelectric material for capacitors and the development of new electro-optical materials.

"This project placed CVC in a much better position within the industry and led to the turnaround of CVC's research and development efforts in thin film process solutions for a number of applications, such as aluminum oxide and copper, that have become commercially successful for CVC," comments Zoe Piliero, Director of Corporate Communications.

NASA applications for a thin film deposition system rested on production of high performance microwave system components, including power dividers, combiners, switched filters, and phase shifters utilizing superconducting material films.

CVC developed a unique sputter deposition system capable of depositing thin films of high temperature conducting material over relatively large areas.

Work on the deposition system set in motion sales of over $1 million, further strengthening the company's ability to provide deposition services. Assisted by the JPL SBIR support, CVC has become a leader in creating products and thin film process solutions.

The earlier SC-4000 sputter deposition system had five basic modules: the pumping system, baseplate, process chamber, control center, and load lock. A virtually unlimited combination of options and accessories was available for users of the deposition system. The system was equipped with high temperature heat, substrate rotation, thermal evaporation sources, ion gun,

Specialist prepares CVC's system for depositing superconducting thin films.

electron beam guns, and auto process controls. A range of oxides, selenides, and silicides was used for resistance, semiconductor, and photoconductive thin films.

For sputtering, the CVC hardware featured a microcomputer-based quartz crystal monitor that displayed thickness and deposition rates in real time. In an evaporation process mode, the quartz crystal monitor sensor was located within the evaporation area. One sensor was supplied with each evaporation source.

The sputter deposition system developed by CVC became an important tool for researchers who were developing ferroelectric capacitors and new electro-optical components for the communication systems of the future. In addition, the system enabled the use of high temperature superconducting materials in microwave communication systems, improving their efficiency and performance. High temperature superconductor films for microwave devices are ideal, as microwave losses are far below that of copper.

A worldwide supplier of thin film process equipment for semiconductor integrated circuits and thin film magnetic recording head manufacturing, CVC's annual revenues more than doubled in fiscal year 1996 over 1995, highlighted by the company's growth in the Asian markets, with 1997 also a banner year.

CVC's current thin film processing system, CONNEXION®, currently ranges in price from $1.5 million to $3 million, depending on the number of modules and features selected with the system configuration. CVC has an installed base of over 260 process modules on more than 80 CONNEXION cluster tool platforms.

Recording head, semiconductor, and flat panel display technologies continue to evolve day by day. CVC's goal is to continually invest in research and development and to improve equipment and process performance—placing them as a world leader by providing customers with the very best thin film process solutions.

®CONNEXION is a registered trademark of CVC, Incorporated.

Substrate heating unit is made of inconel. High temperatures reached by the heater are key to depositing superconducting thin films.

POWER PLAY—LASER STYLE

Exciting research in the areas of physical chemistry, biochemistry, and atomic physics can be studied utilizing the first commercially available single-frequency laser diode system.

SDL, Inc., San Jose, California, is a leader in the design, manufacture, and sale of semiconductor lasers, optoelectronic integrated circuits, and system and fiber optic-related products.

With partial support from the Ames Research Center's (ARC's) Small Business Innovation Research (SBIR) program, SDL has demonstrated and started marketing the first high-power, single-frequency continuously tunable laser diode system suitable for high-precision spectroscopy. High power gives the laser unprecedented attributes. Tapping more than a half watt of power, the SDL TC40 laser operates in the mid-infrared spectral regions. With frequency doubling, laser light in the UV and visible regions can be generated. It has several advantages over other mid-infrared lasers, including high reliability, lower power consumption, room temperature operation, and compact size.

The NASA connection to the laser's creation is part geological, part biological, and part atmospheric. Mid-infrared laser sources are of interest because many geologically and biologically important gases have strong absorption lines in this spectral region. Absorption is the process in which radiant energy is retained by a substance.

Absorption spectroscopy can perform simple, direct measurements of gas concentrations. Measurement of the ratio of carbon dioxide gases is particularly important for tracing material sources and studying biological activity.

Sometime in the 21st century, human crews will adventure to Mars. Those future missions are likely to include onboard mid-infrared laser sources for analysis of Mars rock and soil and atmosphere.

Mid-infrared spectroscopy has several additional NASA applications, including the monitoring of gas concentrations in spacecraft cabins, measuring atmospheric gas concentrations, and analyzing meteorites.

Commercial applications for mid-infrared gas sensing include pollution monitoring of factory and automobile emissions.

"With complete electronic control of scanning, power and modulation, in amplitude as well as frequency, and no optical adjustments, the laser can be put to work within minutes of its arrival," says Milan Zeman, SDL's Product Marketing Manager for instrument and medical systems. By greatly reducing both the equipment and personnel costs of entry, the tunable laser enables researchers with limited budgets to pursue important scientific study, Zeman adds.

Part of the Ames SBIR support led to laser prototypes for high resolution spectroscopy applications. Spectroscopic measurements throughout the visible to mid infrared can be used for process control in semiconductor manufacturing, where monitoring of moisture content, metallizations, epitaxial depositions, and optical coatings can increase device yields. Other industrial chemical processes could be monitored by the laser as well.

SDL's products are focused on two principal markets. The first is composed of applications where optoelectronic solutions are replacing electronic-based systems. The second market uses optoelectronic products as replacement light sources for conventional optical technologies. SDL is focusing on an array of telecom, cable television, networking and satellite communications applications that predominantly fall into those two categories.

The building of high-power, single-frequency continuously tunable laser diodes will enable further expansion of new spectroscopic applications. By offering performance that was previously available only from tabletop-sized complex laser systems, this laser diode technology is expected to bring powerful analytical techniques out of development laboratories and into widespread commercial use.

High-power, single-frequency continuously tunable commercial laser diode system. The compact system opens up research prospects for studies in biological and atmospheric processes, and for pollution monitoring.

BREAKING THE WAVELENGTH BARRIER

Compact, inexpensive lasers are of growing interest because of a diverse suite of applications in global climate monitoring, as well as medical diagnostics and process control. Many organic compounds, including hazardous effluents, have characteristic absorption spectra in the wavelength region of 2-4 microns. But attaining these wavelengths in convenient room-temperature diode lasers was considered a tough, technological challenge.

New Focus, Inc., Santa Clara, California, has been busy at work, tackling a new frontier of laser technology. Company work in the area has been aided by Small Business Innovation Research (SBIR) awards from Goddard Space Flight Center. SBIR funds have melded the talents of New Focus, its research and development subsidiary, Focused Research, Inc. (FRI), also of Santa Clara, and with subcontractor Sarnoff Corporation of Princeton, New Jersey.

FRI does the research for innovative products that require a longer development time than most New Focus products. FRI partners with leading research groups in universities, industry, and government to work on the leading edge of optical technology. New Focus provides the ultimate link to commercialization through its manufacturing and marketing skills in the photonics community. FRI's objectives are to maximize the return on research and development investments by providing low cost and innovative solutions to problems for government and commercial customers.

FRI and New Focus, working with Sarnoff, have developed single-mode tunable diode lasers with center wavelengths from 1.9 microns to 2.3 microns. As an offshoot from this work, New Focus was able to introduce the 2.0 micron external cavity diode laser into their commercial product line. These lasers are expected to be particularly useful in applications such as medical diagnostic experiments, trace gas detection, environmental sensing, and optical seeding of lidar (laser radar) systems. Advancing these state-of-the-art laser systems has also driven advances in materials engineering and anti-reflection coatings.

New Focus has set out to become a leading supplier of photonics research tools. The company has six product lines: tunable lasers, high performance photodetectors and receivers, laser modulators, electro-optic instruments, mechanical positioners, and optical components. The company has established a solid presence and reputation in the marketplace. New Focus has commercialized a suite of ultra-narrow linewidth diode lasers that are tunable over a broad range of frequencies. The lasers are user-friendly, reliable, and reasonably priced. An ongoing objective of New Focus is developing products with advanced capabilities at low cost.

For the government, its interests are in rapid technology development and commercialization with with concern that future suppliers who are capable, interested, and have the infrastructure to achieve low-cost manufacturing are available.

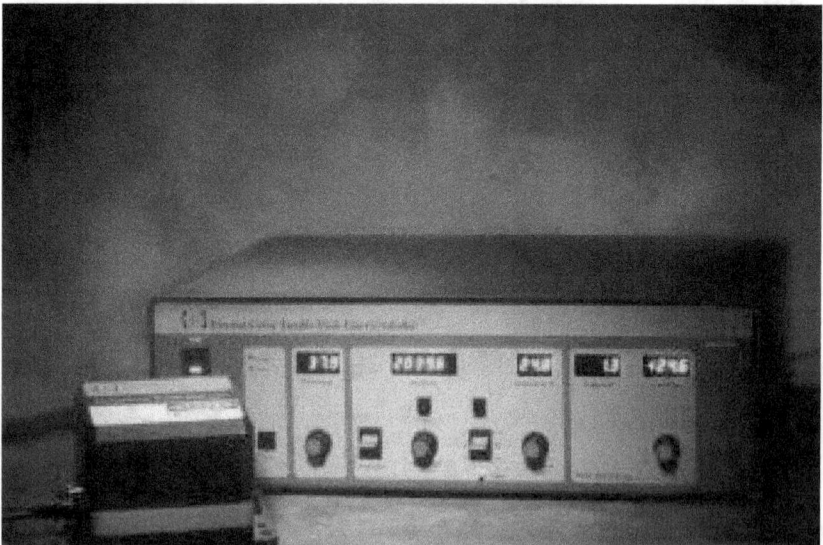

A custom-designed 2.0 micron diode laser has been developed, a level of capability that should prove commercially important, as well as valuable to the scientific community.

SEAL THE DEAL

Adding more reliability to electronic parts is always of foremost interest to high-tech manufacturers. A two-year study has mastered a process technology to render plastic encapsulated electronics modules more resistant to moisture and environmental contaminants.

AvanTeco, Whittier, California, and Revtek, Inc., of Torrance, California, now produce, market, and sell a newly-devised coating method for electronic packaging. This technology was brought to fruition through Small Business Innovation Research (SBIR) contracts from the Jet Propulsion Laboratory (JPL). This novel work was initially performed by Engineering Technologies Associates of Laguna Hills, California.

Considered risky is the packaging of electronic parts in low-cost plastics. While fine for most commercial devices, plastic packaging of critical electronic circuits within expensive spacecraft and medical equipment required further reliability studies. The approach taken by researchers was to originate an overcoat for circuits with an inorganic moisture barrier. A low temperature vacuum deposition process that was non-damaging to underlying epoxy and the assembled circuit proved a workable solution. This overcoating reduced moisture absorption of encapsulated samples. Also, the dual coating had low weight loss in thermal vacuum and low toxicity level as tested to NASA specifications.

Over 90 percent of all integrated circuits produced today are either transfer molded or liquid encapsulated in epoxy plastic. Multimillions of these plastic-packaged chips are produced each day for the consumer and commercial electronics markets. They perform reliably for the intended environments. However, with the emergence of large, very high density, high speed chips, new concerns about the use of plastic encapsulation have been raised.

Making this issue more debatable is the drive to reduce costs of civilian and military hardware, an approach that calls for use of commercial off-the-shelf electronics, or COTS for short. Therefore, enhancing circuit reliability through use of protective coatings, instead of hermetically-packed integrated circuits, has become increasingly suggested. But there are several problems to this approach, one being the exposure of plastic-packaged parts to high humidity at extreme temperatures. Plastics, no matter how good, are permeable to moisture over time, which can then penetrate to the interior of the device and cause a catastrophic failure.

JPL-supported research teams evaluated candidate organic and inorganic barrier coatings and chemically inert plastics. The winning combination? Applying an

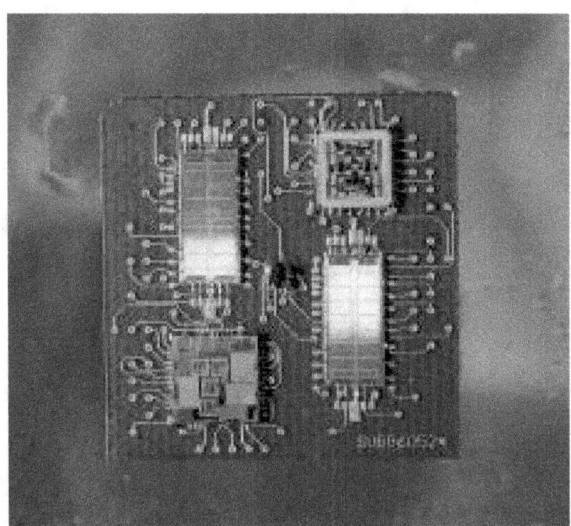

High-value electronic circuitry can be overcoated with an inorganic moisture barrier. Shown is a circuit board, replete with an alert chip (center) that signals if moisture has penetrated the circuit surface.

inorganic barrier coating over the plastic, a technique that greatly reduces moisture penetration and improves reliability. A process was then developed for overcoating epoxy encapsulated circuits with a secondary inorganic dielectric film for added moisture protection. To increase long-term moisture resistance, adhesion between coatings was optimized.

The dual coating approach is expected to add moisture protection to commercial epoxy-coated parts at low added cost, and to reduce packaging cost for many high reliability circuits for space, military, and medical use.

SBIR contract work also yielded a low-cost and easily tested moisture and contamination sensor chip. It acts as a "moisture fuse," says Dr. James Licari, President of AvanTeco. "Once moisture has completely penetrated to the active surfaces of a plastic encapsulated module, the device will short out," he says. "However, using this failure mechanism in a beneficial way detects the penetration of moisture and ionic contaminants through the plastic or into a hermetically sealed enclosure. The failure of the chip is an alert that moisture has penetrated the circuit surface and that the circuit is at risk due to corrosion or other moisture-induced failure," Licari says. The moisture sensor chips were developed jointly by AvanTeco and Revtek, and are being produced and sold by Revtek.

LASER CRYSTAL

VLOC Incorporated of New Port Richey, Florida, has carved out a unique place in the laser industry. A diversity of product coupled with vertical integration of the manufacturing process has made this possible. Exploration by the company in certain laser materials was kick-started by NASA Small Business Innovation Research (SBIR) awards.

Langley Research Center supported VLOC through an SBIR award to carry out crystal growth and materials characterization research. That support led to an exquisite laser crystal, now being commercialized with the firm's own internal funding.

VLOC is a subsidiary of II-VI Incorporated of Saxonburg, Pennsylvania. NASA SBIR work was initiated through Lightning Optical Corporation, based at Tarpon Springs, Florida, which was acquired by II-VI Incorporated in 1996, and was merged with II-VI's Virgo Optics facility in July 1997 to form VLOC, Incorporated.

VLOC grows and manufactures oxide and fluoride laser grain crystals as well as various nonlinear materials. A customer catalog offers specialty optics, cavities, and components for lasers that operate from the ultraviolet to the near infrared region of the spectrum.

The ultimate result of the Langley-funded SBIR work is the commercial availability in the marketplace of a reliable source of high-quality, damage resistant laser material, primarily for diode-pumping applications: Chromium-doped Lithium Strontium-Aluminum-Fluoride (Cr:LiSAF) crystals.

This laser material was sought by NASA for employment in a solid-state diode pumped laser that would be extremely compact. This first Cr-activated, directly diode pumped laser was required to be broadly tunable in the near infrared region of the spectrum, highly efficient, yet straightforward in its design. LiSAF is well-suited for the generation and amplification of pulses as short as a few tens of femtoseconds.

High performance Cr:LiSAF materials can feature several different high reflector, anti-reflection and dichroic optical coatings, depending on the type of laser that uses the laser crystal.

Several immediate tasks for Cr:LiSAF crystals have been projected. Commercial interest in remote sensing of the atmosphere, to determine water vapor, as well as pollution monitoring, and noninvasive surgical techniques, indicates a multitude of applications and markets exist for a device incorporating VLOC-grown and fabricated Cr:LiSAF. The company is interacting with four laser companies to provide Cr:LiSAF crystals for future commercial sales.

VLOC anticipates annual sales of the LiSAF product lines to exceed a quarter million dollars in the first year following completion of the Langley Research Center-awarded Phase II SBIR funding.

As a subsidiary of II-VI Incorporated, VLOC manufactures virtually all of the optical components required for solid-state lasers which includes optics, glass cavities, and crystals. The quality of crystal products is a result of VLOC's tight control of the process, from starting materials though crystal growth, fabrication, coating, and verification. VLOC uses state-of-the-art coating techniques, such as ion-assisted ion beam sputtering, which stands up to the rigors of high power laser use on a production basis. Most of the company's new, purpose-built facilities on the Gulf Coast of Florida are dedicated to manufacturing.

In 1997, revenues and earnings growth set record levels at II-VI, explains Francis Kramer, President and Chief Operating Officer of the company. "Expanded use of lasers and the continued development of radiation detection technology increased and diversified the markets we serve," he says.

II-VI Incorporated and VLOC are on a pathway for continued growth and expansion. From manufacturing to health care to automotive industries, a promising future is anticipated in fabricating specialty materials that enable lasers and other devices to operate at peak efficiency. "As we move forward, our focus remains on expanding our core capabilities while identifying opportunities for acquiring new businesses and technologies to continue our leadership position," Kramer explains.

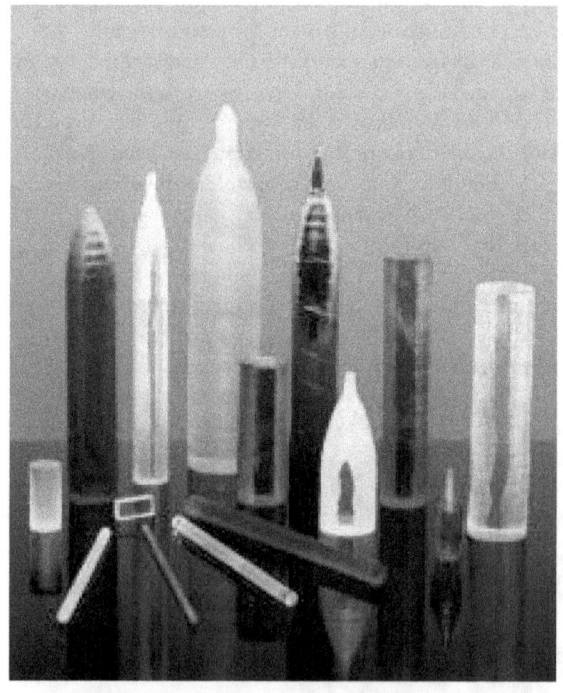

Custom manufacturing of laser crystals and optical components satisfy expanding industrial, medical, environmental, and research markets.

ALL-IN-ONE CHIP

Under a Lewis Research Center Small Business Innovation Research (SBIR) contract, EMC Technology, Cherry Hill, New Jersey, established a family of temperature-compensating electronic attenuators. These devices reduce the impact of heat upon amplifier gain.

Special thermistor materials were developed by EMC Technology, thanks to the SBIR-funded work, devices that are particularly useful in certain types of low power amplifiers used in satellite applications. This novel approach to temperature compensation spurred the company's Thermopad™ temperature compensating attenuator product line.

Not only did the materials developed satisfy NASA requirements, the SBIR work has also proven useful on several new commercial fronts. Devices with greater temperature compensation and lower loss have been developed and used by companies such as Hughes Space and Communications, Motorola, Lucent, Ericcson, and General Instrument.

One added result stemming from the EMC Technology work has been a component that provides a temperature-compensated DC voltage that is proportional to the power dissipated in a radio frequency termination with a frequency range of 200 to 6000 Megahertz. Communication systems that require accurate, reliable, low cost power detection for level control and alarm circuits have benefited by this development.

This peerless set of products was enhanced by materials developed with the funding from an SBIR Phase II award, says Joseph Mazzochette, EMC Technology's Vice President of Engineering. "The performance improvements due to these materials is quite dramatic," he says. Moreover, as an added company dividend, those enhancements open the door to new applications.

For over three decades, EMC Technology has been an industry leader with innovative, high quality microwave components. EMC Technology products are a mainstay in the wireless telecommunications industry, in particular. Knowing that the needs of the marketplace change quickly, the company is poised to respond to these changes with cost-effective new solutions.

The passive temperature compensating attenuator best examples this company strategy.

Now patented by EMC Technology, this component is an absorptive microwave attenuator, providing power dissipation that varies with temperature. The device can be used in any application that requires a known amount of attenuation change for a particular temperature shift. This is particularly useful for maintaining the output of gain stages, mixers, power dividers, and other signal processing components over temperature.

This electronic component is the ideal temperature compensation solution, claims EMC Technology, for cost, performance, and reliability. Presently used closed-loop temperature compensation circuits can be replaced with a single chip device requiring no bias or control. It excels in multiple signal applications such as cellular telephone networks. With low cost and no signal distortion, the attenuator provides high reliability for spacecraft applications. Available in a wide assortment of package styles, the compensating attenuator reduces component count, increases reliability, and saves the buyer money.

™Thermopad is a trademark of EMC Technology.

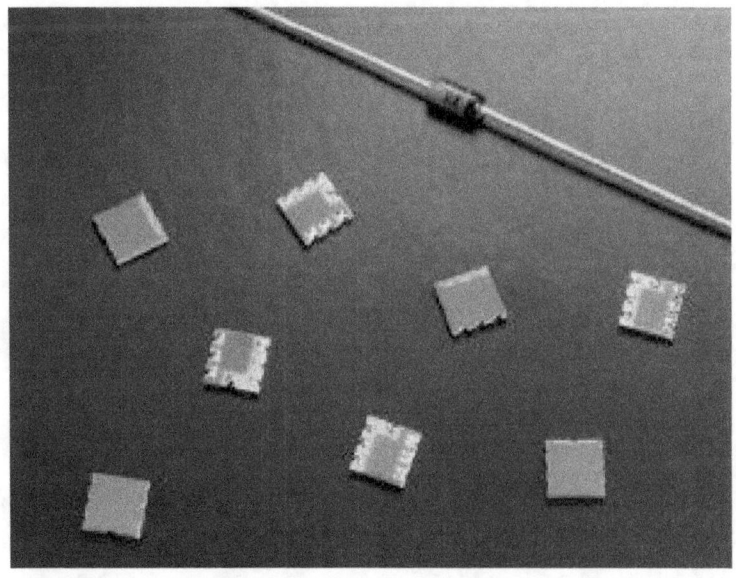

NASA-funded materials research helped in the fabrication of electronic attenuators that compensate for temperature changes in amplifiers, mixers, and oscillators.

NASA Success and Education—A Special Feature

This year, Spinoff highlights two special programs at the National Aeronautics and Space Administration. By also helping to improve the competitiveness of our nation in the world community, these programs complement the efforts of NASA's Commercial Technology Program.

First, an overview of the assistance given by NASA to U.S. businesses. NASA has established the Technology Outreach Program to provide technical assistance to local businesses. The program applies scientific and engineering innovations originally developed for space applications to technical problems experienced by companies. Through the support of its own research laboratories, NASA has solved technical problems of businesses of all sizes and varieties from making ink dry faster in the manufacture of American flags to improving the fit of a prosthetic foot.

Second, as a government agency whose basic product is the advancement of human knowledge, NASA hopes that the inspiration and intellectual excitement inherent in the Aeronautics and Space Program will enrich the study of social sciences, life sciences, physical sciences, mathematics, and technology at all levels of education. NASA is committed to promoting excellence in education, supporting the teaching profession, and increasing awareness of the impact science, mathematics, and technology will have on the quality of life in the 21st century. Providing access and engagement in NASA's exciting missions is key to NASA's education vision of promoting excellence in America's education systems.

PARTNERSHIPS FOR PROGRESS

Never has the future seemed so close at hand.

In just a few year's time, the globe enters a new millennium—one that is filled with exciting promise and, assuredly, difficult challenges. The early decades of the 21st century will test America's competitive prowess in the global marketplace as much as before, as well as the nation's entrepreneurial verve.

NASA stands ready to partner with U.S. businesses, to help solve design challenges, and to improve the overall productivity and competitiveness of American industry.

This cooperation is a partnership for progress.

There are numbers of examples of how NASA outreach to a major company, or the smallest of private business, has been of benefit to the Nation as a whole. These are success stories.

For instance, the Kennedy Space Center (KSC) and the State of Florida have a technology outreach program—one that helps Florida business solve technical problems. A myriad of problem-solving ideas have come forward, from making ink dry faster in the manufacturing of American flags, and improving the fit of a prosthetic foot, to identifying a low-cost, hand-held carbon monoxide detector for fire fighters, and even reducing the noise and vibration of a dental drill.

The Technological Research and Development Authority (TRDA) was established in 1987 by the Florida Legislature. KSC and TRDA, working together, have been able to transfer space technology expertise to private industry needs. A Rockledge, Florida company that makes high-speed electric motors was plagued by bearing breakdown at high speed. NASA identified an improved bearing design and materials that would result in longer-lasting performance. In another case, a Cape Coral, Florida manufacturer of an oxygen device for home use had a problem on its equipment. KSC engineers tested the device, identified the source of the problem, and initiated the solution. NASA's KSC has come to the assistance of hundreds and hundreds of Florida companies over the past few years.

Ames Research Center technology was put into place at the Denver International Airport (DIA). The largest airport constructed in the past two decades, DIA features the Center Terminal RAdar CONtrol Automation System (TRACON)—an air traffic control automation system developed at Ames in the late 1980s. TRACON is the Denver International Airport's primary traffic management tool. Denver airport officials say the airport's air traffic management systems have been operating very smoothly and efficiently since the airport began running with TRACON.

Another NASA success story involved Jennifer, a young woman diagnosed with Leigh's Encephalopathy. Her only method for communicating was a single-switch interface with a computer. She activated the device with her mouth, allowing her to select from an array of words. However, the switch under her chin was cumbersome and difficult to use. Her specific needs were addressed by Langley Research Center at the request of SkillQuest Services, of the city of Virginia Beach. Langley engineers devised a custom headpiece for the youngster. A "chin mouse" was crafted so that the girl's chin could trip a microswitch, allowing her to select an option on a computer screen. As a result, this specially-crafted device has improved Jennifer's quality of life.

Lewis Research Center's Space Communications Division, the Cleveland Clinic, and the University of Virginia are working jointly under a NASA Space Act Agreement. Satellite transmission of mammograms through NASA's Advanced Communications Technology Satellite (ACTS) has been possible, with images beamed to the location of mammography experts. Now patients in rural, urban, and low population density regions, as well as economically depressed areas can have experts review their mammograms. ACTS allows the images to be delivered in near real time, thanks to the satellite's high data rate and high-tech ground station network.

NASA's Commercial Remote Sensing Program (CRSP) at the John C. Stennis Space Center worked with a realtor to jointly develop a better way for prospective home buyers to find a new house. The answer was in looking to the sky. Making use of primarily airborne sensors, CRSP specialists culled together a computer mapping system. Specific geographical data were then referenced to the imagery: Possible flood areas, percent of shade on the lot, setback distance between the street and the house, stores, developments, etc., were all defined. The end product helped the realty company make home buying easier.

Marshall Space Flight Center structures and dynamics engineers helped improve the obstetrical forceps used to properly position an infant in the mother's womb prior to delivery. For this task, the NASA technicians found that by using composite materials, a safe load distribution on the infant could be held in check as pressure was applied by the attending physician. The key was in using fiber optic sensors embedded in the composite material during manufacturing. These sensors enable the physician to monitor pressure on the infant throughout the delivery. These space-age forceps make infant delivery safer.

Lastly, not all NASA success stories can be so fruitful. In one case, it was really the pits—literally.

The Jet Propulsion Laboratory (JPL) was asked by a baking corporation to help devise a way to detect any cherry pits, or pieces of pits, that remain in cherry pie filling. The answer involved use of infrared camera technology, typically used for planetary, Earth, and astrophysical research. The JPL solution now prevents tooth damage to a bakery customer, and subsequent law suits.

Linking NASA to problem solving is a natural. But those problems don't have to be hundreds of miles above Earth or on a faraway world. NASA's technological expertise is not confined to space...it can be further leveraged by building bridges and commercial partnerships between the space agency and the private sector.

NASA is an investment in America's future...and can offer a rich bounty of competence, skills, facilities, and a willingness to create partnerships for progress.

EDUCATIONAL OUTREACH—OPENING UP THE MIND'S EYE

It is called the "Information Age." This revolution throughout the 1990s is founded on a simple truism: Knowledge is power.

From far-flung locales, be it in the home or classroom, students have access to the latest NASA findings and observations. NASA is working with teachers and others in the academic community to inspire America's students and create an array of learning opportunities. The space agency is committed to help enlighten inquisitive minds and involve teachers and students in endeavors that seek answers to fundamental questions of research and science.

This search for educational resources beyond today's traditional learning centers is crucial. The nation must stay on the winning side of the observation made by writer, H.G. Wells: "Human history becomes more and more a race between education and catastrophe". The Strategic Plan for Education was developed to guide the nine NASA field centers and the Jet Propulsion Laboratory. Science, mathematics, and technology education programs and activities are now underway at elementary, secondary, undergraduate, and post doctoral levels. While impossible to adequately portray all the initiatives, the following projects highlight the types of programs that exemplify NASA's resolve to strengthen the educational foundation of the country.

Available through the powerful Internet, connections to over 13,000 files of space-related material via NASA Spacelink is maintained by the Marshall Space Flight Center. This aeronautics and space resource for educators has been up and running since 1988. Among its features are text-only sites that provide a user with the option of speed without graphics, making it compatible with older technology and offering greater accessibility for the visually impaired. A special search engine permits faster searches, providing "results ranking," "natural language" querying, and the ability to search hundreds of other NASA sites on the Internet in addition to the Spacelink library.

Goddard Space Flight Center is creating a low-cost weather station for schools, tuned to geostationary weather satellites. A prototype system has been installed at a neighboring high school. Once trial-tested, an entire network of schools can be interlinked via low-cost satellite receiving stations, enhancing the applicability of using real-time satellite weather data for academic use in schools. Such inter-school links can promote project-oriented learning. Students can learn about meteorological science, computers, teamwork, meeting schedules and objectives, and documenting results.

The NASA-funded Classroom of the Future (COTF) at Wheeling Jesuit College, Wheeling, West Virginia, is providing technology-based tools and information to kindergarten through twelfth grade (K-12) students. COTF's primary mission is to harness advanced computer, video, and telecommunications technologies to help enhance the learning and teaching process for math, science, and technology. COTF is exploring a range of multipurpose, multimedia educational tools for use in teacher training and has established itself as a leader in the field.

Working with the Jet Propulsion Laboratory (JPL), undergraduate and high school students can get an incredible view of the Earth when a Space Shuttle mission is underway via the KidSat mission operations center at the University of California at San Diego (UCSD). While a Space Shuttle is orbiting the Earth, a KidSat mission operations team electronically targets a digital still-camera aboard the space plane, snapping images of selected spots on Earth. These images are then relayed to the KidSat mission control at UCSD, operated by middle school students, and posted on the Internet for students to study and analyze. The Institute for the Academic Advancement of Youth (IAAY) at Johns Hopkins University is developing the KidSat curriculum for middle school students and teachers. The JPL-UCSD-IAAY partnership has proven picture-perfect, with students learning about weather phenomenon, geography, patterns of rivers and irrigation fields.

Electronic field trips for students and teachers were organized by Ames Research Center as part of the JASON Project. A JASON expedition to Yellowstone National Park and Iceland was broadcast via satellite linkup to twenty-seven other locations across the country, with students operating a live video camera at the sites. During the live hook-ups, students could join scientists and engineers as they explored hot springs, geysers, volcanoes, and glaciers, asking questions of the experts during the science treks. Ames researchers discussed with students the link between studying these Earth sites and searching for signs of life on distant Mars.

Kennedy Space Center (KSC) engineers and scientists became virtual guest instructors at twenty southwestern Florida middle schools through an interactive science education program through the Internet. A Virtual Science Mentor program, originating from KSC, used the latest in desktop video technology to bring the mentors and science class students together. Under the program, forty KSC mentors and a similar number of middle school teachers were paired to establish topics to be covered, including physics, astronomy, technology transfer, space engineering, and other subjects.

The Affordable Technology to Link America's Schools (ATLAS) program is a commercialization outgrowth of Langley Research Center research within the High Performance Computing and Communications/Information Infrastructure Technology and Applications (HPCC/IITA) Program. NASA and ATLAS share a joint goal: To connect more than 70,000 K-12 school sites by the year 2000. Working with the Virginia State Housing and Urban Development Office, Langley is to provide ATLAS orientation sessions to explain the technology design and applications of Internet access and Intranet services.

Computers originally used by engineers and researchers at NASA's Lewis Research Center have been donated to students attending schools in the Cleveland Empowerment Zone. Strengthening a student's ability to compete in a highly-competitive world is a clear motivation for the program. Some 100 computers from the NASA center were distributed among ten schools in the Empowerment Zone, providing technology necessary to educate young minds for the 21st century.

The Stennis Space Center (SSC) education programs and activities include TREND 2000-The Technology, Research, Education, and Discovery Lab. This effort facilitates the integration of technology into the curriculum by providing innovative and creative classroom strategies using state-of-the-art technology. Another project fused SSC geographic mapping software with the re-creation of Amelia Earhart's aviation quest, giving students and teachers a computer-generated way to study changes in the terrain, and growth of urban areas from Earhart's flight in 1937 to 1997.

Johnson Space Center (JSC) hosts a variety of educational initiatives, providing first-hand teacher outreach programs that relay the latest news on human space flight projects, from the Space Shuttle to the International Space Station. JSC supports an active cooperative student work program, giving college students valuable experience in technical and business fields by working at the center. JSC participates directly with Space Center Houston, a neighboring complex built to handle public and educator interest in NASA and the American human space flight program.

Students Sharing NASA engages high schools students with research activities at the Dryden Flight Research Center. Students in the program collect information about a NASA research project, then use it to develop their own Internet site on the World Wide Web. Teacher enhancement, undergraduate and graduate programs, as well as postdoctoral programs are available at Dryden.

These brief descriptions in no way reflect the total number and scope of NASA-supported educational initiatives. Hundreds of elementary, secondary, and higher education activities and programs constitute the true NASA commitment to improving the knowledge of students and teachers alike.

In the 1990s, America's educational system is being challenged. Exciting and innovative approaches are underway at NASA that can help improve the competitiveness of the United States in the world community.

NASA's Commercial Technology Network

The NASA Commercial Technology Network (NCTN) extends from coast to coast. For specific information concerning commercial technology activities described below, contact the appropriate personnel at the facilities listed or go to the Internet at **http://nctn.hq.nasa.gov**. General inquiries may be forwarded to the National Technology Transfer Center.

To publish your NASA spinoff story, contact the NASA Center for AeroSpace Information or go to the Internet at **http://www.sti.nasa.gov/tto/contributor.html**.

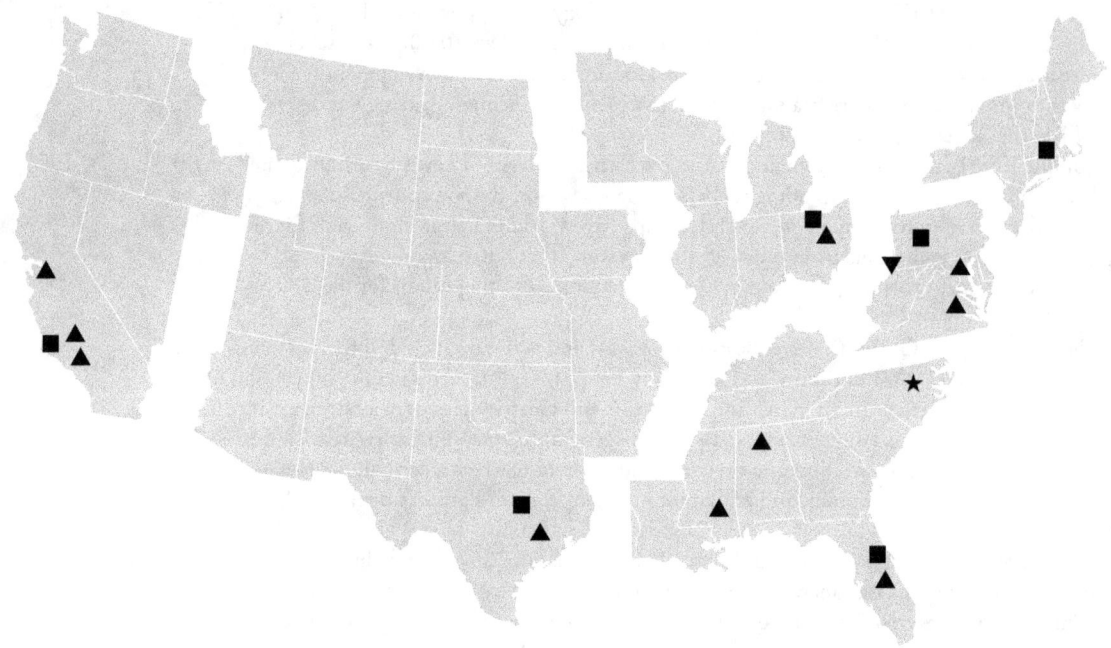

▲ **Field Center Commercial Technology Offices**
Represent NASA's technology sources and manage center participation in technology transfer activities

▼ **National Technology Transfer Center**
Provides national information, referral, and commercialization service for NASA and other government laboratories

■ **Regional Technology Transfer Centers**
Provide rapid access to information, as well as technical and commercialization services

★ **Application Team**
Provides a range of technology management services including technology assessment, valuation and marketing; market analysis; intellectual property audits; commercialization planning; and the development of partnerships.

▲ FIELD CENTERS

Ames Research Center
National Aeronautics and
    Space Administration
Moffett Field, California 94035
Chief (acting), Commercial Technology Office:
Caroline Blake
Phone: (650) 604-6646
email: cblake@mail.arc.nasa.gov

Goddard Space Flight Center
National Aeronautics and
    Space Administration
Greenbelt, Maryland 20771
Manager, Commercial Technology Office:
George Alcorn, Ph.D.
Phone: (301) 286-5810
email: galcorn@pop700.gsfc.nasa.gov

Lyndon B. Johnson Space Center
National Aeronautics and
    Space Administration
Houston, Texas 77058
Director, Technology Transfer and
Commercialization Office:
Henry Davis
Phone: (281) 483-0474
email: henry.l.davis@jsc.nasa.gov

John F. Kennedy Space Center
National Aeronautics and
    Space Administration
Kennedy Space Center, Florida 32899
Associate Director, Technology Programs and
Commercialization Office:
James A. Aliberti
Phone: (407) 867-6224
email: jim.aliberti-1@kmail.ksc.nasa.gov

Langley Research Center
National Aeronautics and
    Space Administration
Hampton, Virginia 23681-0001
Director, Technology Applications Group:
Joseph S. Heyman, Ph.D.
Phone: (757) 864-6005
email: j.s.heyman@larc.nasa.gov

Lewis Research Center
National Aeronautics and
    Space Administration
21000 Brookpark Road
Cleveland, Ohio 44135
Chief, Commercial Technology Office:
Larry Viterna
Phone: (216) 433-3484
email: larry.a.viterna@lerc.nasa.gov

George C. Marshall Space Flight Center
National Aeronautics and
    Space Administration
Marshall Space Flight Center, Alabama 35812
Director, Technology Transfer Office:
Sally A. Little
Phone: (205) 544-4266
email: sally.a.little@msfc.nasa.gov

Jet Propulsion Laboratory
4800 Oak Grove Drive
Pasadena, California 91109
Manager, Commercial Technology Program Office:
Merle McKenzie
Phone: (818) 354-2577
email: merle.mckenzie@jpl.nasa.gov

NASA Management Office—JPL
4800 Oak Grove Drive
Pasadena, California 91109
Technology Commercialization Officer:
Arif Husain
Phone: (818) 354-4862
email: ahusain@nmo.jpl.nasa.gov

John C. Stennis Space Center
National Aeronautics and
    Space Administration
Stennis Space Center, Mississippi 39529
Manager, Technology Transfer Office:
Kirk V. Sharp
Phone: (228) 688-1914
email: kirk.sharp@ssc.nasa.gov

Dryden Flight Research Center
National Aeronautics and
    Space Administration
Post Office Box 273
Edwards, California 93523-0273
Chief, Public Affairs, Commercialization and
    Education Office:
Lee Duke
Phone: (805) 258-3802
email: duke@louie.dfrc.nasa.gov

▼ NATIONAL TECHNOLOGY TRANSFER CENTER

Wheeling Jesuit University
Wheeling, West Virginia 26003
Joseph P. Allen, president
Phone: (304) 243-2455
email: jallen@nttc.edu

■ REGIONAL TECHNOLOGY TRANSFER CENTERS

1-800-472-6785
You will be connected to the RTTC in your geographical region.

Far-West
Technology Transfer Center
University of Southern California
3716 South Hope Street, Suite 200
Los Angeles, California 90007-4344
Kenneth E. Dozier, Jr., director
Phone: (213) 743-2353
email: kdozier@bcf.usc.edu

Northeast
Center for Technology Commercialization, Inc.
1400 Computer Drive
Westborough, Massachusetts 01581
William Gasko, Ph.D., director
Phone: (508) 870-0042
email: wgasko@ctc.org

Mid-West
Great Lakes Industrial Technology Center
25000 Great Northern Corp. Ctr., Suite 260
Cleveland, Ohio 44070-5320
Christopher Coburn, executive director
Phone: (216) 734-0094
email: coburnc@battelle.org

Southeast
Southern Technology Application Center
University of Florida
College of Engineering
Box 24
One Progress Boulevard
Alachua, Florida 32615-9987
J. Ronald Thornton, director
Phone: (904) 462-3913
email: jrthorn@nervm.nerdc.ufl.edu

Mid-Continent
Texas Engineering Extension Service
Texas A&M University System
301 Tarrow Street, Suite 119
College Station, Texas 77843-8000
Gary Sera, director
Phone: (409) 845-8762
email: ecsera@teexnet.tamu.edu

Mid-Atlantic
University of Pittsburgh
3400 Forbes Avenue, 5th Floor
Pittsburgh, Pennsylvania 15260
Lani Hummel, executive director
Phone: (412) 383-2500
email: lhummel@mtac.pitt.edu

★ TECHNOLOGY APPLICATION TEAM

Research Triangle Institute
Post Office Box 12194
Research Triangle Park, North Carolina 27709
Doris Rouse, Ph.D., director
Phone: (919) 541-6980
email: rouse@rti.org

NASA CENTER FOR AEROSPACE INFORMATION

Spinoff Project Office
NASA Center for AeroSpace Information
7121 Standard Drive
Hanover, Maryland 21076-1320
Walter Heiland, manager
Phone: (301) 621-0241
email: wheiland@sti.nasa.gov

Spinoff Development Team

Project Manager:
Walter Heiland

Graphic Design:
John Jones

Sr. Editor/Writer:
Amy Harding

Photography:
Kevin Wilson

Editor:
Danielle Israel

Writer:
Leonard David

Editorial Support:
Deborah Drumheller

Project Support:
Doris Wahl

www.ingramcontent.com/pod-product-compliance
Lightning Source LLC
Chambersburg PA
CBHW081728170526
45167CB00009B/3745